SpringerBriefs in Geography

SpringerBriefs in Geography presents concise summaries of cutting-edge research and practical applications across the fields of physical, environmental and human geography. It publishes compact refereed monographs under the editorial supervision of an international advisory board with the aim to publish 8 to 12 weeks after acceptance. Volumes are compact, 50 to 125 pages, with a clear focus. The series covers a range of content from professional to academic such as: timely reports of state-of-the art analytical techniques, bridges between new research results, snapshots of hot and/or emerging topics, elaborated thesis, literature reviews, and in-depth case studies.

The scope of the series spans the entire field of geography, with a view to significantly advance research. The character of the series is international and multidisciplinary and will include research areas such as: GIS/cartography, remote sensing, geographical education, geospatial analysis, techniques and modeling, landscape/regional and urban planning, economic geography, housing and the built environment, and quantitative geography. Volumes in this series may analyze past, present and/or future trends, as well as their determinants and consequences. Both solicited and unsolicited manuscripts are considered for publication in this series.

SpringerBriefs in Geography will be of interest to a wide range of individuals with interests in physical, environmental and human geography as well as for researchers from allied disciplines.

Radmila Salic Makreska · Zoran Milutinovic

Seismic Hazard Zonation for the Republic of North Macedonia

Springer

Radmila Salic Makreska
Department for Risk, Disaster Management and Strategic Planning, Institute of Earthquake Engineering and Engineering Seismology (IZIIS)
Ss. Cyril and Methodius University
Skopje, North Macedonia

Zoran Milutinovic
Department for Risk, Disaster Management and Strategic Planning, Institute of Earthquake Engineering and Engineering Seismology (IZIIS)
Ss. Cyril and Methodius University
Skopje, North Macedonia

ISSN 2211-4165 ISSN 2211-4173 (electronic)
SpringerBriefs in Geography
ISBN 978-3-032-13944-3 ISBN 978-3-032-13945-0 (eBook)
https://doi.org/10.1007/978-3-032-13945-0

This Springer imprint is published by the registered company Springer Nature Switzerland AG
The registered company address is: Gewerbestrasse 11, 6330 Cham, Switzerland

Preface I

This monograph is the result of many years of work by a team of researchers and collaborators from IZIIS. Its primary aim is to trace the development and progress of scientific and professional achievements in the field of seismic hazard, with a particular focus on the territory of present-day North Macedonia, viewed through a historical lens. As a fundamental component and initial step in seismic risk analysis, the reliable assessment of seismic hazard is essential. Given the significant efforts made in this domain, both scientifically and in practice, we, as authors, felt a responsibility to collect, systematically analyze, compare, and present these contributions to the broader scientific and professional community.

This monograph is published in memory of Dr. Zoran Milutinovic, a respected professor and colleague, who remained actively engaged with the team until his final days. His dedication greatly enriched the quality and depth of this work. With the publication of this monograph, the remaining team of researchers and close collaborators express their deep gratitude for his lasting contributions to the field, his guidance and professional leadership, and his role in advancing seismic hazard assessment.

Skopje, North Macedonia — Radmila Salic Makreska

Preface II

The contribution presents a systematic and comprehensive review of the major seismic hazard assessment and zonation accomplishments for the territory of the present Republic of North Macedonia from 1948 to 2024. This contribution encompasses all significant efforts related to territory of North Macedonia on national, regional, and global levels of seismic hazard assessments, briefly discussing the methodologies used and the results derived. The monograph compactly presents and preserves historical, administrative, and some technical information that is currently of limited availability. It identifies and emphasizes the adverse influence of the migration of design seismicity enforced by different seismic zonings and non-harmonized standards for seismic design, concrete, and reinforce concrete, on a timescale.

The monograph presents, in necessary detail, the spatial and temporal evolution of seismic zoning maps during the last 76 years, the results of efforts by the Macedonian national and international research community to quantify the Macedonian seismic environment, and the implications thereof on modeling the spatial and temporal evolution of developed exposure.

The authors considered it worthwhile to collect, systemize, analyze, and make available for further studies and research to the broader scientific and professional community. The presented review may be of particular interest to research communities of countries succeeding the Socialist Federal Republic of Yugoslavia (SFRY) since the discussed legal technical documents have controlled the development of the built environment (exposure) and its seismic vulnerability in the pre-Eurocode design and construction period, not only in the territory of the Republic of North Macedonia but also in the territories of all SFRY successors.

All maps and studies presented in this monograph have been processed using GIS and/or digitized from the original maps.

Keywords Seismic hazard · Exposure · Vulnerability · Safety · Eurocodes · Republic of North Macedonia

Skopje, North Macedonia

Radmila Salic Makreska
Zoran Milutinovic

Acknowledgements The authors would like to express their sincere gratitude to the Institute of Earthquake Engineering and Engineering Seismology (IZIIS), Ss. Cyril and Methodius University in Skopje, for its continuous institutional support, resources, and longstanding commitment to the advancement of seismic hazard assessment in the Republic of North Macedonia.

Special thanks are extended to all domestic and international authors whose scientific contributions in the field of seismic hazard have significantly enriched the knowledge base and served as a foundation for this monograph.

We would like to express our appreciation to our young collaborators, Zabedin Neziri, Daniel Tomic, Jovan Trajcevski, and Mirko Dimitrovski, for their dedicated support in GIS processing, data digitization, and technical tasks throughout the development of this monograph.

We are especially grateful to Dr. Laurentiu Danciu for his continuous support to the advancement of seismic hazard modeling in North Macedonia and for providing access to the most recent European research and data.

This publication also underwent a formal review process at the Ss. Cyril and Methodius University in Skopje and received positive evaluations from five reviewers, whose expertise and constructive feedback greatly improved the quality of the work. We gratefully acknowledge the contributions of Profs. Ana Trombeva Gavriloska, Igor Pesevski, Katerina Drogreska, Marta Stojmanovska, and Julijana Bojadzieva. Their collective review was published in the University Bulletin (in Macedonian), issue No. 1311, dated 15.07.2024.

This monograph is the result of a collective scientific and professional effort, shaped by decades of cooperation, knowledge-sharing, and a shared commitment to seismic risk reduction.

Competing Interests The authors have no competing interests to declare that are relevant to the content of this manuscript.

Contents

About the Authors

Dr. Radmila Salic Makreska is Full Prof. at the Institute of Earthquake Engineering and Engineering Seismology, Ss. Cyril and Methodius University in Skopje (North Macedonia). She received her master's degree in Earthquake Engineering and a Ph.D. in Technical Sciences at Ss. Cyril and Methodius University in Skopje (North Macedonia). Her activities are mainly focused on seismic hazard and risk analysis, disaster management, and strategic planning studies. She has broad scientific and professional experience in participating and leading national and international projects. Her professional activities are also part of several consultancy projects for different international organizations and institutions such as UNDP, WHO, WB, EBRD, JICA, and others. She is Member of several crisis and disaster management national bodies established under the mandated governmental institutions, as well as several professional and scientific associations. At present she is General Secretary of the Macedonian Association for Earthquake Engineering (MAEE). Her contributions to the field are evident through her publication record, which includes over 60 scientific papers presented at national and international conferences, more than 120 professional reports, and more than 15 publications in reputable journals. She is Member of editorial boards of several international journals. Dr. Salic Makreska is one of the authors of the current national seismic zoning maps part of NA for EC8. Recently she was elected as ExeCom Member of European Facilities for Earthquake Hazard and Risk (EFEHR). She has received several awards for her scientific and professional work.

Dr. Zoran Milutinovic (1950–2023) was Full Professor at the Institute of Earthquake Engineering and Engineering Seismology, Ss. Cyril and Methodius University in Skopje (North Macedonia). He was Esteemed Scholar in earthquake engineering, disaster management, and strategic planning. He received his Ph.D. from Kyoto University in Japan. He authored and co-authored over 500 scientific publications and two books and served on editorial boards of several international journals. He was Professor at IZIIS from 1992 and Visiting Professor at the Universities of Kobe, Kyoto, and Belgrade. He managed the Department for Risk, Disaster Management, and Strategic Planning at IZIIS until 2017 and led the European Center for the Vulnerability of Industrial and Vital Systems (ECILS) from 1997 to 2018. He managed numerous national and international projects and served as Consultant for different UN agencies and the Council of Europe. Prof. Milutinovic was General Secretary of the Macedonian Association for Earthquake Engineering (MAEE), National Delegate in European (EAEE) and World (IAEE) Associations for Earthquake Engineering, and Member of the Executive Board of the NGO Balkan Security Forum. He received the Civil Protection Cross Rank III from the Bolivarian Republic of Venezuela and numerous awards for his scientific and professional work, particularly in post-earthquake missions, scientific assignments, and strategic planning. Prof. Dr. Zoran Milutinovic is Leading Author of the current national seismic zoning maps part of NA for EC8.

Abbreviations

AREC	The Agency for Real Estate Cadastre of RNM
BALKANS-OQ	Regional Seismic Hazard and Risk Modeling Initiative Project
BSHAP	Harmonization of Seismic Hazard Maps for the Western Balkan Countries Project
CEN	European Committee for Standardization
CRC	Concrete and Reinforce Concrete
DLR	Damage Limitation Requirement
EC	Eurocode
EFEHR	European Facilities for Earthquake Hazard and Risk
ELF	Equivalent Linear Force
EN	European Norm
EPOS	European Plate Observing System
ESHM13	European seismic Hazard Model 2013
EU	European Union
FEMA	US Federal Emergency Management Agency
FNRY	Federal People's Republic of Yugoslavia
GEM	Global Earthquake Model
GIS	Geographic Information System
GMM	Ground Motion Model
GMPE	Ground Motion Prediction Equation
GPS	Global Positioning System
GSHAP	Global Seismic Hazard Assessment Program
ISRSM	Standardization Institute of RNM
IZIIS	Institute of Earthquake Engineering and Engineering Seismology
JUS	Yugoslavian standard
M	Magnitude
MCS	Mercalli-Cancani-Sieberg Intensity Scale
MRS	Modal Response Spectra
MSK-64	Medvedev–Sponheuer–Karnik Intensity Scale
NATO	North Atlantic Treaty Organization

NCR	No Collapse Requirement
NDSHA	Neo-Deterministic Seismic Hazard Assessment
OGoRM	Official Gazette of Republic Macedonia
OGoRNM	Official Gazette of the Republic of North Macedonia
OGoSFRY	Official Gazette of SFRY
OQ	Open Quake
PGA	Peak Ground Acceleration
PSHA	Probabilistic Seismic Hazard Assessment
RISK-UE	Advanced Approach to Earthquake Risk Scenarios With Application to Different European Towns Project
RM	Republic of Macedonia
RNM	Republic of North Macedonia
RP	Return Period
SERA	Seismology and Earthquake Engineering Research Infrastructure Alliance for Europe Project
SESAME	Seismotectonic and Seismic Hazard Assessment of the Mediterranean Basin Project
SFRY	Socialist Federal Republic of Yugoslavia
SHARE	Seismic Hazard Harmonization in Europe
SNiP	Russian (and former USSR) Technical Standards
SZM	Seismic zoning map
UHS	Uniform Hazard Spectrum
UKIM	Ss. Cyril and Methodius University in Skopje, North Macedonia
UNISDR	UN International Strategy for Disaster Reduction
MKC	Macedonian Standard

List of Figures

List of Tables

Chapter 1
Introduction

Abstract This chapter introduces the seismotectonic setting and seismic risk context of the Republic of North Macedonia (RNM). Situated in the Western Balkans, RNM is exposed to moderate to high seismic hazard due to ongoing extensional tectonics and has experienced numerous damaging earthquakes, including twelve events with $M_L \geq 6.0$ during the last century. The chapter adopts the UNISDR risk framework and emphasizes that, despite four commonly distinguished risk components (hazard, exposure, vulnerability and capacity), they can be conceptually grouped into the natural environment, which cannot be controlled, and the policy environment, which shapes exposure, vulnerability, and capacity through regulations and standards. It briefly reviews the historical role of seismic design codes, concrete and foundation standards, and their accompanying seismic zoning maps in controlling the seismic safety of the built environment. Finally, the chapter outlines the objectives and organization of the monograph, setting the scene for the systematic review of official regulations, zoning maps, and research studies presented in the subsequent chapters.

Keywords Seismic hazard · Seismic risk · Risk framework · Seismic design codes · Seismic zoning · Exposure and vulnerability · Republic of North Macedonia

The Republic of North Macedonia, hereafter RNM, is situated in the Western Balkan,[1] Southeastern Europe, in a region characterized by moderate-to-high seismic hazard due to ongoing permanent tectonic extensional processes (Dumurdjanov et al. 2005; Fig. 1.1). Only in the last century, RNM has experienced 12 earthquakes with a magnitude of $M_L \geq 6.0$ (Cejkovska et al., 2016; Fig. 1.2), resulting in a significant number of human casualties and economic losses. Given the frequent occurrence of strong seismic activity, the reliable assessment of seismic hazard and risk

[1] Western Balkans is a term used in the European Union to refer to six countries in Southeastern Europe that are covered by EU enlargement policy: Republic of Albania, Bosnia and Herzegovina, Montenegro, Republic of Kosovo, Republic of North Macedonia, and Republic of Serbia.

R. Salic Makreska and Z. Milutinovic, *Seismic Hazard Zonation for the Republic of North Macedonia*, SpringerBriefs in Geography,
https://doi.org/10.1007/978-3-032-13945-0_1

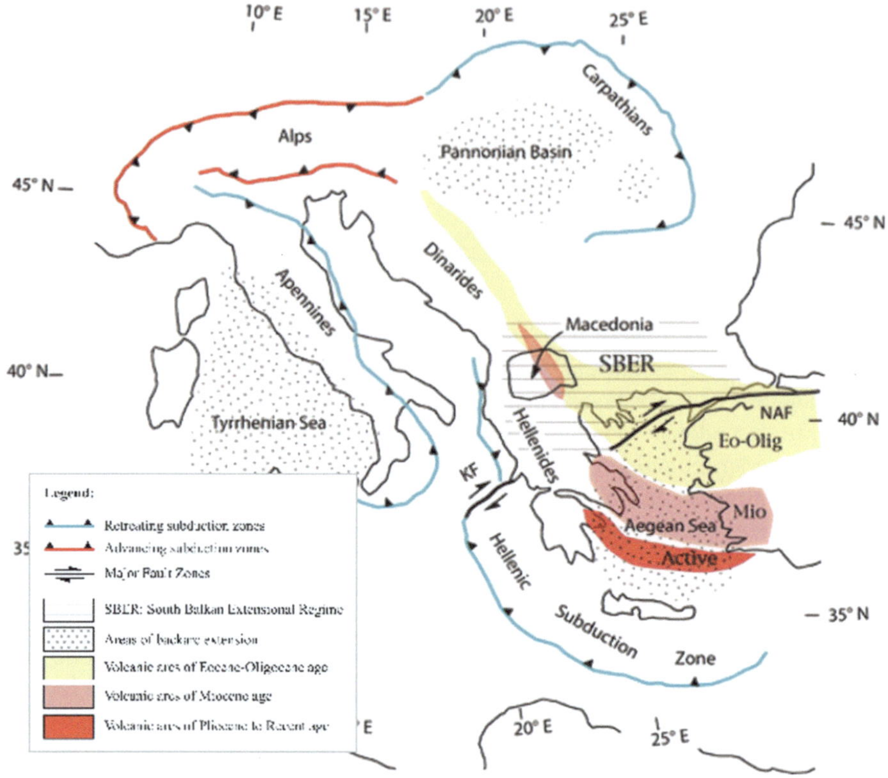

Fig. 1.1 Simplified tectonic map of Eastern Mediterranean region. Dumurdjanov et al. (2005)

remains of paramount interest for RNM, serving as the foundation for elaborating and implementing various earthquake-disaster risk reduction measures and strategies.

Risk is a combination of the probability of an occurrence of a hazardous event and its consequences, which results from interactions between hazard, vulnerability, exposure, and capacity (UNISDR 2004). Risk assessment (and associated risk mapping) includes a review of the technical characteristics of hazard such as its location, intensity, frequency, and probability; the analysis of exposure and its vulnerability, including the physical, social, health, economic and environmental dimensions; and the evaluation of the effectiveness of prevailing and alternative coping capacities in respect to likely risk scenarios (UNISDR 2009) (Fig. 1.3).

While UNISDR distinguishes four elements controlling the risk (Fig. 1.3), theoretically, there are only two: (a) the natural and (b) the policy environments. The occurrence, time of occurrence, and magnitude of adverse natural events are largely or fully (such as earthquakes) uncontrollable by humans and should be mitigated by adequate counter-measure policies. The man-made property (exposure) and its vulnerability, as well as the required capacity, strictly depend on policies, codes, and standards applied during the development of the built environment.

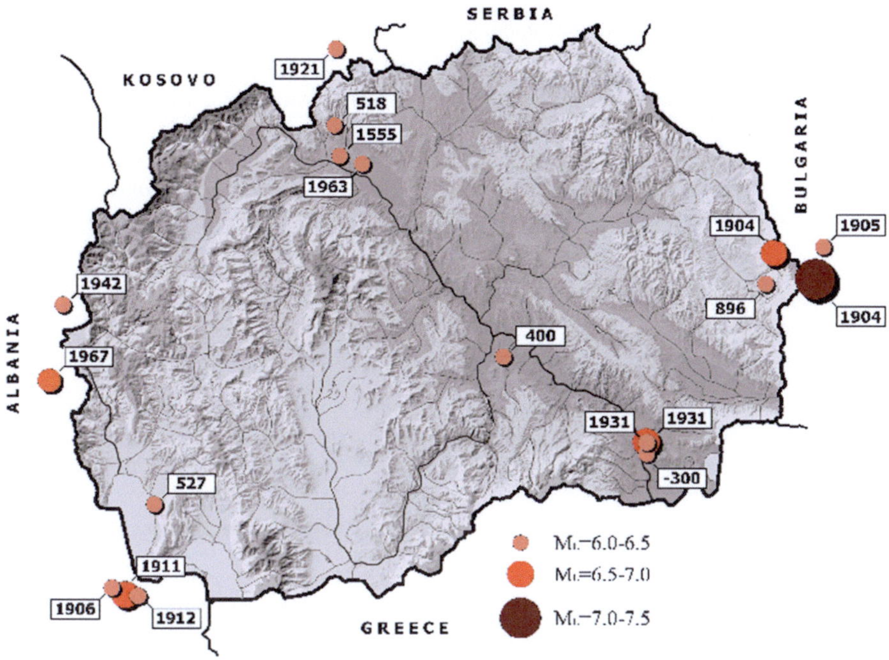

Fig. 1.2 Distribution of earthquakes for RNM and border regions (300BC–2014). *Source* Cejkovska et al. (2016)

Fig. 1.3 Risk elements

Historically, the seismic safety of the construction in RNM has been controlled by, among others, seismic design codes, codes for concrete and reinforced concrete, and foundation codes. Each seismic design code was accompanied by a seismic zoning map prescribing the strength of the seismic design environment (Table 1.1).

The organization of this contribution is as follows: (a) Section 2 discusses and presents the official seismic design regulations and seismic zoning maps; (b) Section 3

Table 1.1 Seismic design codes and accompanying seismic zoning maps for RNM

<table>
<tr><th colspan="2">Building code</th><th colspan="2">Seismic zoning map</th></tr>
<tr><th>Year</th><th>Building code title</th><th>Year</th><th>Seismic zoning map title</th></tr>
<tr><td>1948</td><td>Provisional Technical Regulations for Loading of Structures
Привремени технички прописи за оптерећење зграда (in original)
OGoFNRY No. 61/48 of July 17, 1948</td><td>1948</td><td>Division of Territory of Federal People's Republic of Yugoslavia into Seismological Zones
Podela teritorije Federativne Narodne Republike Jugoslavije u seizmološke zone (in original)
OGoFNRY No. 61/48 of July 17, 1948</td></tr>
<tr><td rowspan="2">1964</td><td rowspan="2">Rulebook on Provisional Technical Regulations for Construction in Seismic Regions
Правилник за привремените технички прописи за градење во сеизмичките подрачја (in original)
OGoSFRY No. 39/64 of September 30, 1964</td><td>1964</td><td>S.F.R. Yugoslavia Seismological Map
S.F.R. Jugoslavia seizmološka karta (in original)
OGoSFRY No. 39/64 of September 30, 1964</td></tr>
<tr><td rowspan="2">1979</td><td rowspan="2">Seismic Zoning Map of S.R. Macedonia
Карта на сеизмичко реонирање на СР Македонија (in original)
OGoSRM No. 2/79 of January 31, 1979</td></tr>
<tr><td rowspan="3">1981</td><td rowspan="3">Rulebook on Technical Regulations for Construction of Buildings in Seismic Regions
Правилник за техничките нормативи за изградба на објекти на високоградба во сеизмички подрачја (in original)
OGoSFRY No. 31/81 of June 5, 1981 (Amendments 49/82, 29/83, 21/88, and 52/90)</td></tr>
<tr><td>1982</td><td>Provisional Seismological Map of SFRY
Привремена сеизмолошка карта на СФРЈ (in original)
OGoSFRY No. 49/82 of August 13, 1982</td></tr>
<tr><td>1990</td><td>The Socialist Federal Republic of Yugoslavia, Seismological Maps for Return periods of 50, 100, 200, 500, 1000, and 10,000 years
Социјалистичка Федеративна Република Југославија, Сеизмолошки карти за повратен период од 50, 100, 200, 500, 1000 и 10,000 години (in original)
OGoSFRY No. 52/90 of September 7, 1990</td></tr>
<tr><td>2020</td><td>Rulebook Amending the Rulebook on Design Standards and Norms
Правилник за дополнување на правиникот за стандарди и нормативи за проектирање (in original)
Eurocodes (EN)
OGoRNM No. 211/20 of September 2, 2020
OGoRNM No. 183/23 of August 31, 2023</td><td>2020</td><td>The Republic of Macedonia, Seismic Zoning Map with Elements of Seismic Hazard for Return Periods of 95 and 475 years
Република Македонија, Карта на сеизмичко зонирање со елементи на сеизмички хазард за повратен период од 95 и 475 години (in original)
МКС EN 1998-1/НА:2020</td></tr>
</table>

discusses and presents the seismic hazard maps developed by domestic and foreign authors; (c) Section 4 presents seismic hazard maps developed under regional and global research studies; (d) Section 5 covers discussion and conclusions; and (e) Section 6 presents way forward.

A comprehensive review of seismic hazard and seismic zoning maps developed under various research studies shows that comparing and assembling them in a single contribution are a step toward qualitatively better modeling of exposure, vulnerability, and seismic risk.

The Agency for Real Estate Cadastre of RNM (AREC, https://ossp.katastar.gov.mk/OSSP/) provided grid data on national, municipal, and cadastral municipalities. Map spatial data were georeferenced and digitized from the best originals. For large projects, data were downloaded from relevant internet sites and/or gathered from IZIIS's archives. In all calculations, the RNM area is defined as 25.713 km^2.

References

Cejkovska V, Pekevski L, Drogreska K, Najdovska J (2016) Report per the project of the standardization Institute of the Republic of Macedonia entitled National Annexes for Eurocodes. Faculty of Natural Sciences and Mathematics, Seismological Observatory, Ss. Cyril and Methodius University, Skopje February 2016 (in Macedonian)

Dumurdjanov N, Serafimovski T, Burchfiel CB (2005) Cenozoic tectonics of Macedonia and its relation to the South Balkan extensional regime. Geosphere 1:1–22. https://doi.org/10.1130/GES00006.1

Rulebook Amending the Rulebook on Design Standards and Norms, OGoRNM No. 211/20 of 2.11.2020

UNISDR (2004) Living with risk, 2004 UNISDR terminology on disaster risk reduction

UNISDR (2009) UNISDR terminology on disaster risk reduction. International Strategy for Disaster Reduction, United Nations

Chapter 2
Official Seismic Design Regulations and Zoning Maps

Abstract This chapter provides a comprehensive overview of the official seismic design regulations and associated seismic zoning maps that have governed construction in Republic of North Macedonia and the former SFRY over the last 76 years. It first summarizes the evolution of seismic design standards from JUS61/48 through JUS39/64 and JUS31/81 to the national adoption of EN 1998-1 (Eurocode 8), highlighting the physical equivalence between the JUS total seismic coefficient and the Eurocode behavior factor q, as well as the common R-format definition of base shear used in both ELF and MRS procedures. The chapter then reviews six generations of official seismic zoning maps (1948, 1964, 1979, 1982, 1990, 2020), describing their underlying catalogs, intensity scales, spatial resolution, and assigned seismic intensities and PGA values for major towns. Particular attention is paid to the transition from macroseismic intensity maps to probabilistic PGA-based zonations and to the 2020 DLR/NCR maps, which introduce cadastral municipality-based zonation consistent with Eurocode 8 requirements.

Keywords Seismic design regulations · JUS standards · Eurocode 8 · Seismic zoning maps · Peak ground acceleration · Intensity zonation · Cadastral municipalities · Republic of North Macedonia

2.1 Official Seismic Design Regulations

The official seismic zonation spans over the period of 76 years, from 1948 till 2024, during which four seismic design regulations were in effect (Table 1.1).

JUS61/48 is relevant for buildings only. It defines the seismic coefficient as "minimum horizontal loading" for three "damage zones." Assigned values are in the range of 1–3% of the seismic weight of the building. Details are provided in OGoFNRY No. 61/48.

JUS31/81 and EN 1998-1 define the Equivalent Linear Force (ELF) base shear, F_b^{CODE}, by the following general expression being the product between the total seismic coefficient (K) and the total seismic mass (M) of the building:

R. Salic Makreska and Z. Milutinovic, *Seismic Hazard Zonation for the Republic of North Macedonia*, SpringerBriefs in Geography,
https://doi.org/10.1007/978-3-032-13945-0_2

Table 2.1 Physical meaning and equivalence between JUS and EN 1998-1 (EC8) coefficients

JUS31/81	EN 1998-1	Description
F_b^{CODE}		Base shear
K		Worldwide and in EN 1998-1 termed seismic coefficient; JUS31/81: Total seismic coefficient
G	M	EN 1998-1: Total seismic mass of the building; JUS31/81: Seismic weight of the building
K_0	γ_I	EN 1998-1: Importance factor; JUS31/81: Coefficient of building category
K_S	n/a	Coefficient of seismic intensity. $K_S = PGA/(D_L \cdot g)$
PGA	n/a	MCS-based peak ground acceleration on "characteristic soil" type ground (in g). For K_S values of 0.1, 0.05, and 0.025 assigned to seismic zones of MCS IX, VIII, and VII, intensity degrees corresponding PGA values of are 0.4 g, 0.2 g, and 0.1 g, respectively.
n/a	a_{gR}	Reference peak ground acceleration on A-type ground (in g)
K_P	n/a	Coefficient of ductility and damping; Also referred to as the structural coefficient
R_P	q	*The energy-dissipation capacity of the structure.* EN 1998-1: Behavior factor; JUS31/81: $R_P = D_L/K_P$
$K_D(T)$	$S_e(T)$	EN 1998-1: Elastic horizontal ground acceleration response spectrum also called "elastic response spectrum"; JUS31/81: Coefficient of dynamic response. JUS39/64: "β curve"
λ=1 Dummy	λ	EN 1998-1: F_b the correction factor, the value of which is equal to: $\lambda = 0.85$ for $T_1 \leq 2T_C$ and the building has more than two storeys, or $\lambda = 1.0$ otherwise. JUS31/81: Dummy parameter in Eq. 2.1 (λ=1)
$\eta_{i,k}$		JUS39/64 and EN 1998-1: Mode and the m_k height position-dependent coefficient of vertical distribution
D_L	n/a	JUS39/64 and JUS31/81: Coefficient of allowable structural damage. It is not explicitly formulated in both JUS's and entered JUSs from CH-8-57 (SNiP57) and SNiP II-1. 12-62 (SNiP62) For building and structures in which residual deformations, cracks, damage to individual elements, etc., can be allowed (residential, public, industrial, agricultural buildings and structures; hydraulic and transport facilities; energy and water supply systems, fire stations, fire extinguishing systems, some communication facilities, etc.)$D_L = 4$

$$F_{\mathrm{b}}^{\mathrm{CODE}} = K \cdot \lambda \cdot M\,. \tag{2.1}$$

Modal Response Spectra (MRS) methods of JUS39/64 and EN 1998-1 define the seismic action acting at mass m_k in the ith mode of vibration:

$$S_{i,k}^{\mathrm{CODE}} = K_i \cdot \eta_{i,k} \cdot m_k\,. \tag{2.2}$$

All four standards formulate the story (m_k) and total (M) seismic masses of the building by including all permanent gravity loads (self-weight and additional dead loads), live (imposed) loads reduced by factor 0.5, and snow. JUS61/48 and JUS31/81 define the coefficient of seismic intensity K_S, Table 2.1. as a quotient between the expected mean squarpeak ground acceleration (PGA, in g, $g = 9.81\ \text{m/s}^2$) and the $D_L \cdot g$ product. D_L is a coefficient that is solely dependent on the level of allowable structural damage. The reference value for buildings is $D_L = 4$. To account for the actual energy-dissipation capacity of different building types, JUS standards modify K_S with structural type-dependent "ductility and damping" coefficient, K_P.

$$K_P K_S = \frac{K_P}{D_L} \text{PGA} = \frac{1}{R_P} \text{PGA}. \tag{2.3}$$

JUS R_P coefficient is physically equivalent to EN 1998-1 behavior factor "q," Table 2.1. Values of JUS31/81 ductility and damping factor (K_P) and the corresponding R_P values are tabulated in Table 2.2. JUS39/64 K_P/R_P values are about the same; however, there are slight differences (wording) in defining the structural typology.

The total, K, and the modal K_i, seismic coefficients are:

$$K = \frac{K_o \cdot K_S \cdot K_D}{R_P}. \tag{2.4}$$

Table 2.2 Values of JUS31/81 ductility and damping factor (K_P) and corresponding R_P values

Structural typology		K_P	R_P
(1)	All modern reinforced-concrete (RC) structures, all-steel structures except those listed under item (2) of this table, and all modern wooden structures, except those listed under item (3) of this table	1.0	4.0
(2)	All reinforced masonry structures and braced still frame structures	1.3	3.1
(3)	Masonry buildings strengthened by vertical RC cerclages; All RC shear wall structures with aspect[a] ratios of each wall i lower than 2 (Article 68); very high and slender low damping structures such as high industrial chimneys, antenna towers, and masts, water-towers and other structures with a fundamental natural period of vibration T $\geq$ 2.0s.	1.6	2.5
(4)	Structures with a flexible ground story or any other story, or for structures where there are dramatic changes in stiffness, or with a sudden change in stiffness, as well as ordinary (URM) masonry structures.	2.0	2.0

[a] h_{wi}/h_{wi}; h_{wi} = height of the wall; l_{wi} = length of the wall i

Table 2.3 SZM1979 allowed design MCS modifications by ground sensitivity classes

Symbol	Sensitivity class	Geo-strata description	Allowed design MCS modification
H	Insensitive	Outcropping rock, compact soils	Reduction of MCS intensity for up to 2 degrees
H_1	Less sensitive	Coarse-grained or sandy clay well-consolidated soils with no presence of water	Reduction of MCS intensity for up to 1 degree
H_2	Sensitive	–	No reduction of MCS intensity is allowed
H_3	Very sensitive	–	Construction is not recommended unless seismic microzonation is performed

In essence, irrespectively of the method of structural analysis (ELF or MRS), both JUS and EN 1998-1 standards define seismic coefficient by using the R-format expression that allows simple, intra-, and inter-codal estimation and comparison of the influences of K_o, K_S, K_D, and R_P factors on the magnitude of total (K) and modal (K_i) seismic coefficients.

Further readings on this subject are presented and elaborated in detail in Milutinovic et al. (2022).

2.2 Official Seismic Zoning Maps

RNM and SFRY have a long history of more than 76 years of implementing official seismic design regulations. Starting from the first official building code up to today, four editions of official building codes were enforced in 1948, 1964, 1981, and 2020. Since September 2020, RNM has been in the four-year (3 + 1) parallel implementation of the new European building design regulations (Eurocodes, EN 1990-1999). Considering this timeline, six official seismic zoning maps (SZMs) or sets of maps were adopted and enforced (1948, 1964, 1979, 1982, 1990, and 2020) (Table 1.1, Fig. 2.1).

2.2.1 SZM 1948

Starting in 1948, part of the first official design code was the seismic zoning map published in OGoFNRY No. 61/48 on July 17, 1948 (Table 1.1, Fig. 2.2). This map was issued by the Seismological Institute in Belgrade, Serbia. SZM1948 divides the territory of FNRY into four zones: the zone of "catastrophic destruction, I_{MCS}=IX-X,"

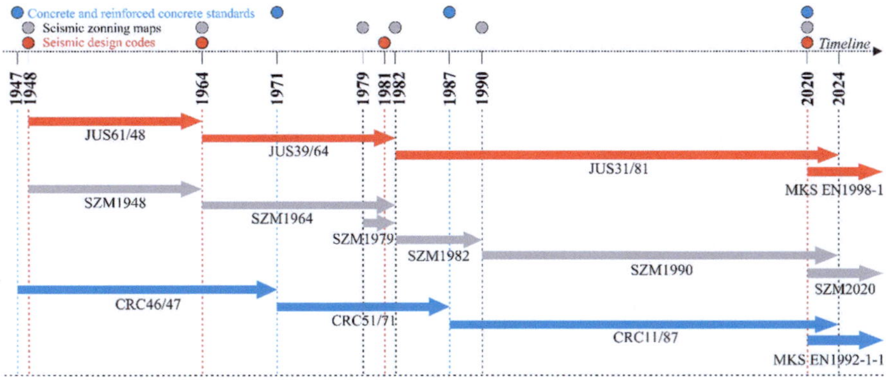

Fig. 2.1 Seismic design codes and related seismic zoning map implementation timeline

the zone with "severe damage, I_{MCS}=VIII," the zone with "minor damage, I_{MCS}=VII," and the zone with "not assigned seismicity, $I_{MCS}\leq$VI."

According to SZM1948, 13.94% of the entire territory of RNM is exposed to IX-X MCS intensity, 30.58% to VIII, 24.06% to VII, and 31.42% have no assigned intensity (VI or less). An extract of the SZM1948 map and the intensities calculated for 15 major towns in RNM, including the city of Skopje, are presented in Fig. 2.2.

2.2.2 *SZM 1964 (1950)*

After two years, in 1950, a new official seismic zoning map was issued by the Seismological Institute in Belgrade, Serbia (Table 1.1, Fig. 2.3). This map was put in effect in 1964. It divides the territory of FNRY into seven intensity zones. Each zone is described by the measure (strength) of earthquake severity and assigned the MCS seismic intensity: the zone "destructive—catastrophic" corresponds to MCS intensity IX, the zone "harmful—significant damage" to VIII MCS, the zone "very strong—minor to moderate damage" to VII MCS, the zone "strong—without damages" to VI MCS, the zone "quite strong—without damage" to V MCS, the zone "moderate" to IV MCS, and the zone "weak, scarcely felt and imperceptible" to III, II, and I MCS. The map is based on an earthquake catalogue comprising all earthquakes affecting the territory of FNRY from 360 to 1950.

As in SZM1948, SZM1964 is characterized by the discontinuity of isolines. Zoning is performed in terms of closed islands characterized by common boundaries. For example, the town of Kichevo, located in the I_{MCS}=VI zone, borders the zones of Debar and Ohrid, both in the I_{MCS}=VIII zone.

According to SZM1964, 22.96% of the total country area belongs to MCS intensity IX, 11.85% to VIII, 22.72% to VII, and 42.47% to intensities $\leq$ VI. The intensities of the major cities are listed in Fig. 2.3.

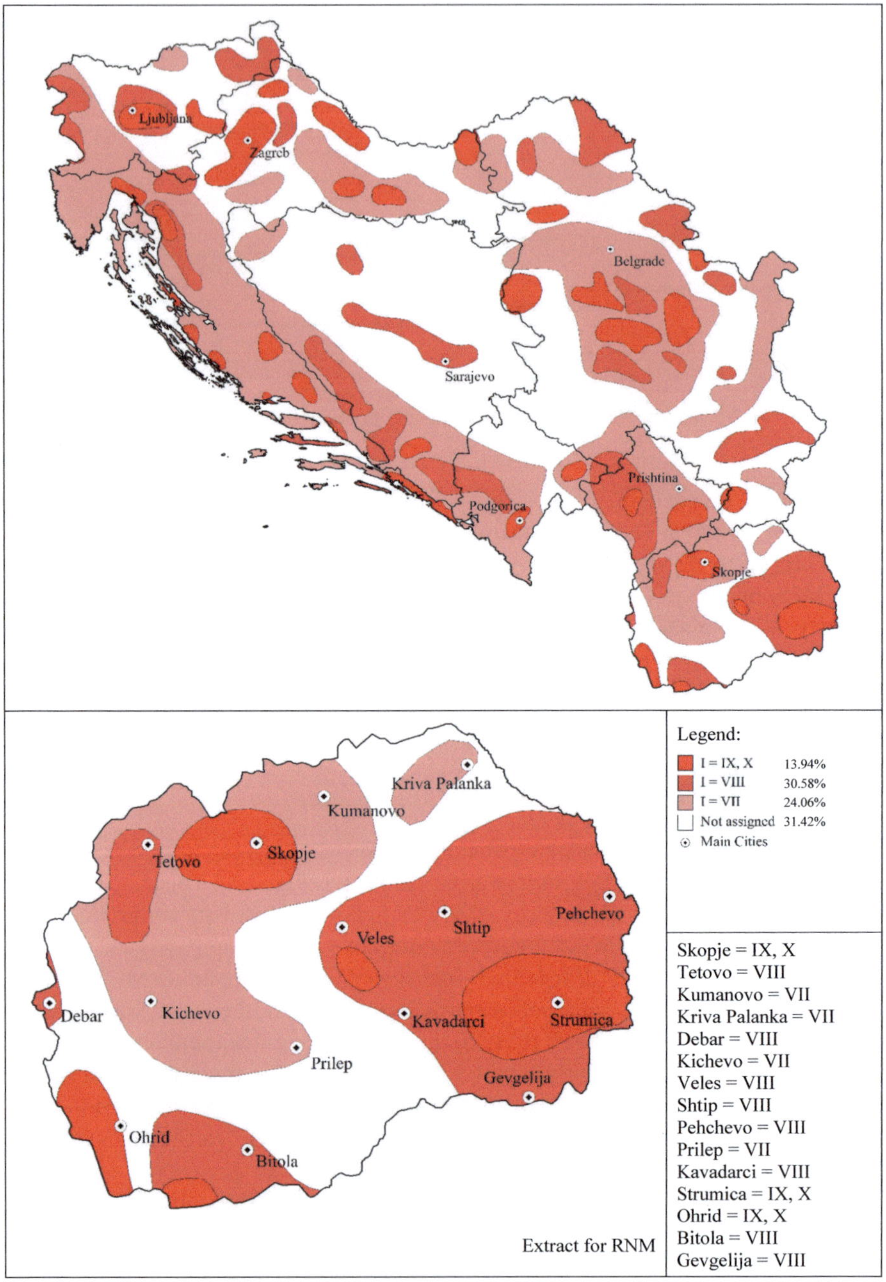

Fig. 2.2 Division of territory of FNRY into seismological zones (SZM1948), adopted in 1948 to accompany JUS61/48 seismic design code

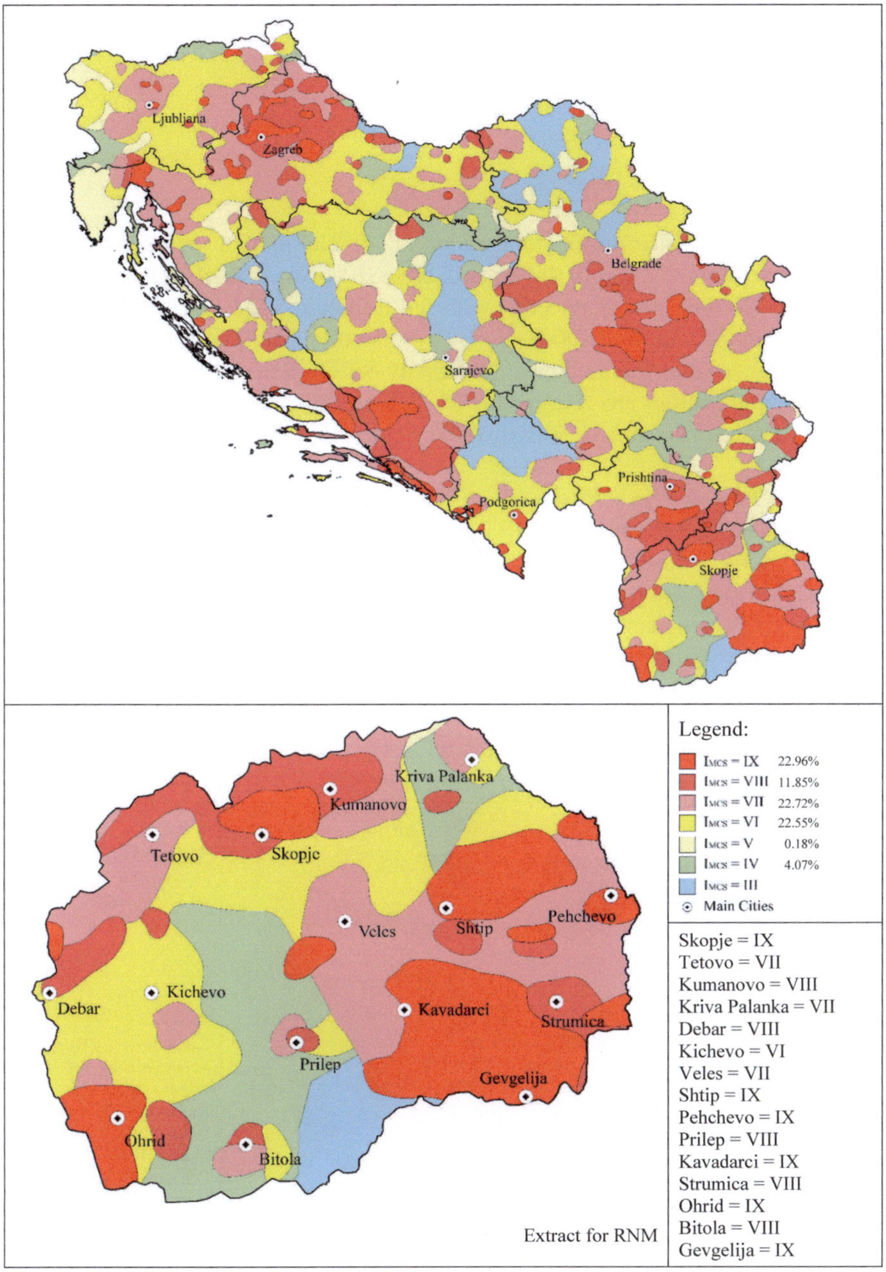

Fig. 2.3 Seismic zoning (seismological) map for SFRY (SZM1950), elaborated in 1950 and adopted in 1964 to accompany JUS39/64 seismic design code

By OGoSFRY No. 39/64 of September 30, SZM1964 was adopted to accompany JUS39/64.

2.2.3 SZM 1979

JUS39/64 allowed federal units to abandon the official federal seismic zoning map and enact their own SZMs. In this regard, OGoSRM No. 2/79 of January 31, 1979, withdrew SZM1964 and replaced it with the Republic SZM (Table 1.1, Fig. 2.4), elaborated on a scale of 1:600,000 by the Seismological Observatory, Faculty of Natural Sciences and Mathematics, University Ss. Cyril and Methodius in Skopje. The mapped intensities are given in the MCS scale with accompanying information about the degree of seismic sensitivity of particular geotechnical strata, ranging from H to H3.

SZM1979 divides the territory of RNM into four seismic zones. MCS intensity X covers 1.94% of the country area, 19.93% corresponds to MCS IX, 53.17% to MCS VIII, and 24.96% to MCS intensity VII. SZM1979 and the intensities estimated for major RNM towns are presented in Fig. 2.4.

A novelty in this map is that it overlaps seismic intensity zones and corrects the design seismicity for regional engineering–geological conditions. SZM1979 recognizes four engineering–geological media and categorizes their sensitivity in four

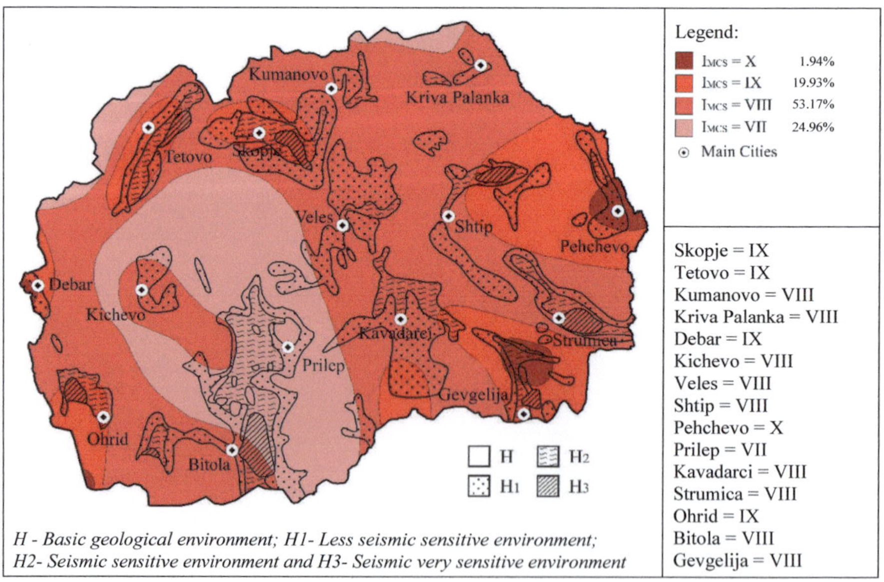

Fig. 2.4 Seismic zoning map of SRM adopted in 1979 to accompany JUS39/64 seismic design code

sensitivity classes, as shown in Table 2.3. The basis for defining these sensitivity classes is the 1967 Engineering Geology Map of SFRY (Cubrilovic et al. 1967), elaborated on a scale of 1:500,000.

Macedonian design practice was very reluctant to implement it because of the small scale (1:600 000) and the poor description and classification of ground conditions incompliant with JUS39/64 ground categorization.

2.2.4 SZM 1982

OGoSFRY No. 49/82 of August 13, 1982 (Table 1.1, Fig. 2.5), enacts a new official seismic zoning map accompanying the JUS31/81 seismic design code. SZM1982 was prepared by the Seismological Institute, Belgrade, Serbia, on a scale of 1:1.000.000. Seismic zoning for the territory of RNM was made by the Seismological Observatory, Faculty of Natural Sciences and Mathematics, Ss Cyril and Methodius University in Skopje. The intensities presented on this map are given in the MCS scale. The map is based on historic earthquake catalogues that cover a period up to 1900 for all $I_{MCS} \geq 7$ earthquakes and all earthquakes with $I_{MCS} \geq 6$ in the period 1900-1978.

SZM1982 divides the territory of RNM into five seismic zones (I_{MCS} VI to X). To I_{MCS} = X zone belongs to 2.55% of the country's territory, 15.88% to I_{MCS}= IX, 32.89% to I_{MCS} = VIII, 42.22% to I_{MCS} = VII, and 6.46% to I_{MCS} =VI intensity. SZM1982 and intensities estimated for major RNM towns are presented in Fig. 2.5.

2.2.5 SZM 1990 (1987)

The Seismological Association of SFRY presented/issued an edition of six probabilistic seismic zoning maps in 1987, scaled at 1:1,000,000. These maps were elaborated, harmonized, and agreed upon by Republic Seismological Surveys (Fig. 2.6). They were expressed in MSK-64 intensities. The Seismological Observatory at the Faculty of Natural Sciences and Mathematics, Ss. Cyril and Methodius University in Skopje, conducted the seismic zoning for the territory of RNM.

For the very first time, the probabilistic seismic hazard methodology was used in the official seismic zoning. The return periods considered (RP) were 50, 100, 200, 500, 1,000, and 10,000 years. The maps were calculated to correspond to a 63% non-exceedance probability (the lifetime of the structure equaled the RP of the map). Reformulating to the zoning standards of nowadays, the RP100 and RP500 maps corresponded to a 9.52% non-exceedance probability in 10 and 50 years, respectively. Both maps are conceptually comparable with EN 1991-1 damage limitation (T_{DLR} = 95 years, or P_{DLR} = 10% in 10 years) and collapse prevention (T_{NCR} = 475 years, or P_{NCR} = 10% in 50 years) requirements. Those probabilistic seismic zoning maps, like all previous SZMs, were based on macroseismic data and were associated with ground conditions referenced as "characteristic ground."

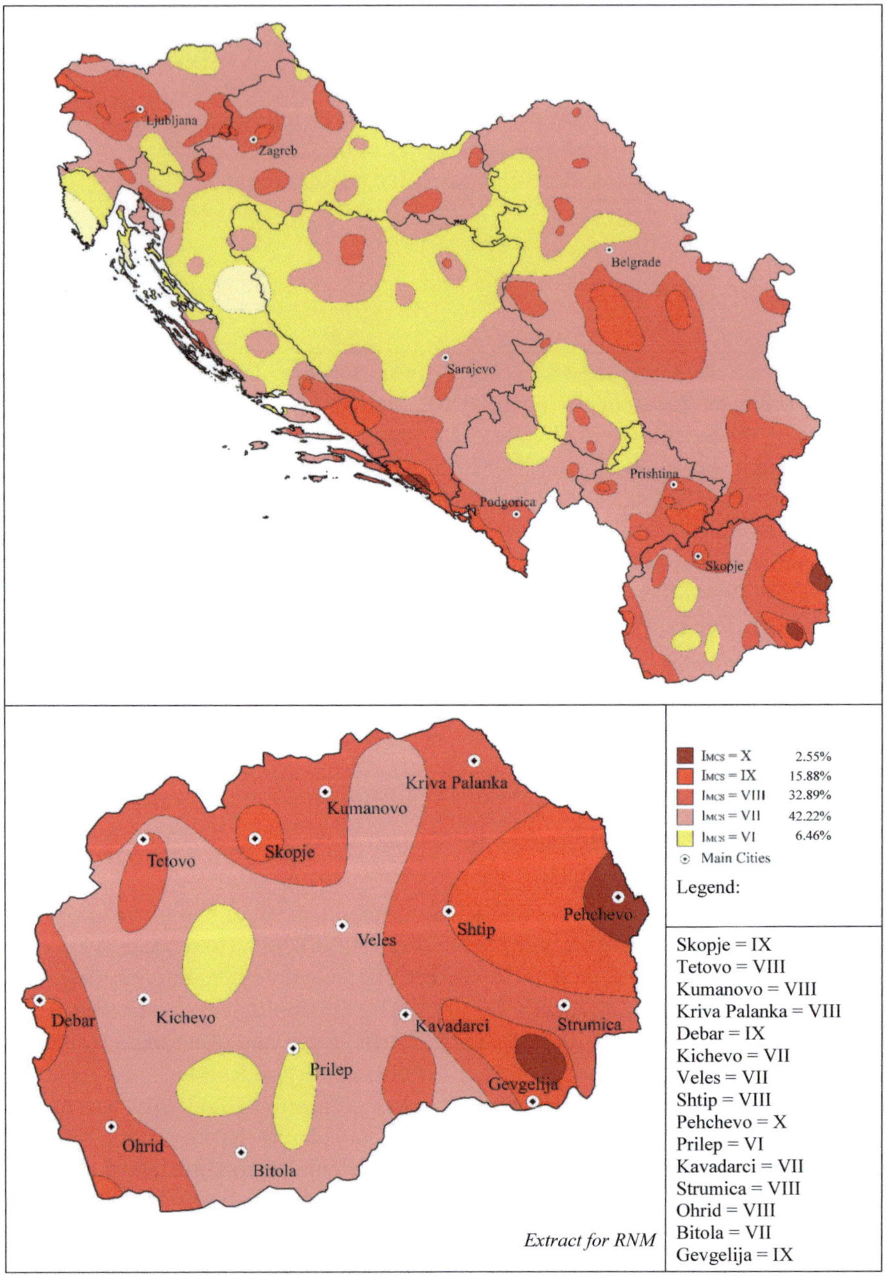

Fig. 2.5 Provisional seismic zoning (seismological) map of SFRY (SZM1982) adopted in 1982 to accompany JUS31/81 seismic design code

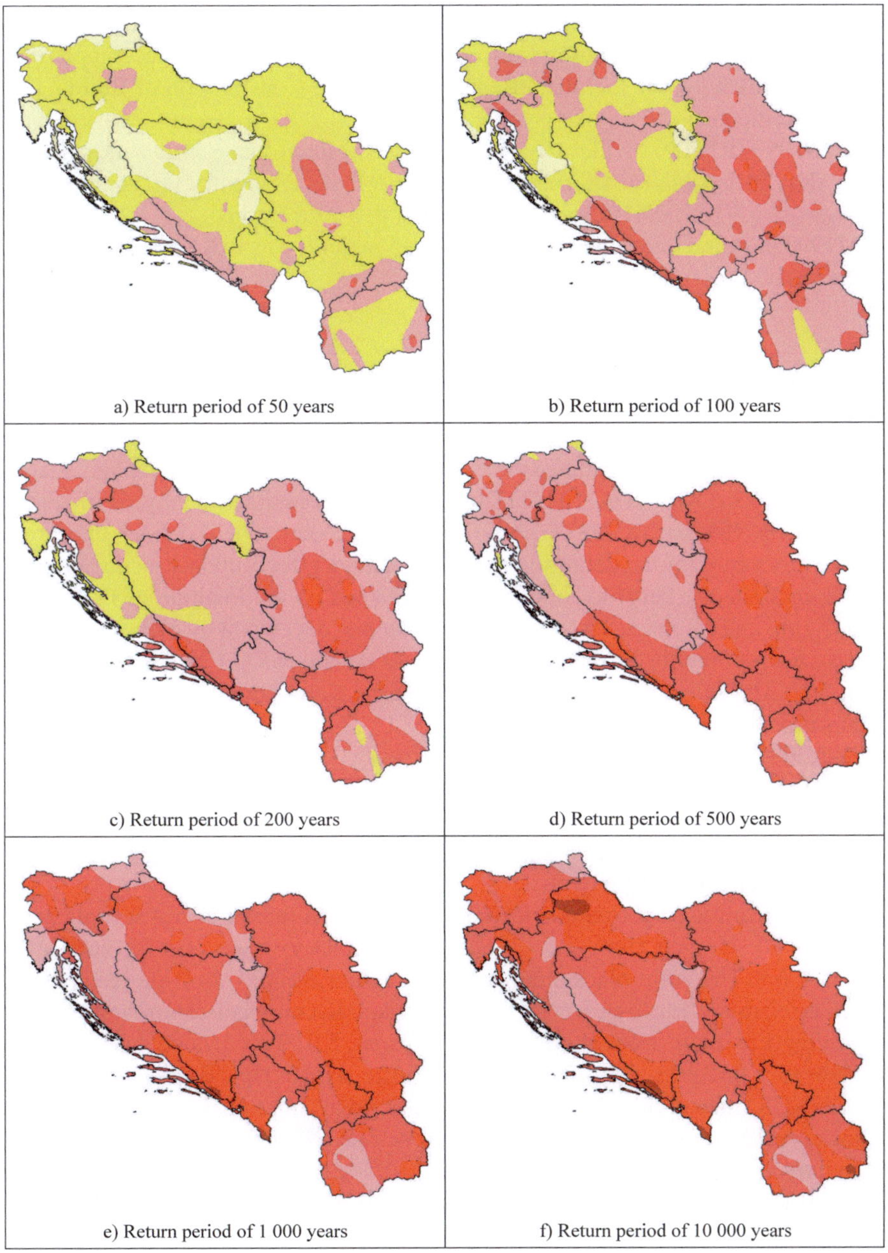

Fig. 2.6 Set of seismic zoning (seismological) maps of SFRY issued in 1987 by the Seismological Association of SFRY for return periods of 50, 100, 200, 500, 1000, and 10,000 years

In 1990, the OGoSFRY No. 52/90 of September 7 (Table 1.1, Fig. 2.7) enforced all six probabilistic SZMs but prescribed the RP 500 years' SZM to accompany the JUS31/81 Rulebook on Technical Regulations for the Construction of Buildings in Seismic Regions.

SZM1990 (RP500) divides the territory of RNM into four seismic zones ($I_{MSK\text{-}64}$ VI to IX). To $I_{MSK\text{-}64}$ = IX zone belongs 5.67% of the country territory, 69.86% to $I_{MSK\text{-}64}$ = VIII, 22.41% to $I_{MSK\text{-}64}$ = VII, and 2.06% to $I_{MSK\text{-}64}$ =VI intensity. SZM1990 map and intensities calculated for 15 major towns in RNM, city of Skopje included, are presented in Fig. 2.7.

2.2.6 *SZM 2020 (2016)*

In September 2020, RNM adopted the Eurocodes for a 3-year parallel implementation with JUS31/81 (OGoRNM No. 211/20), extended for an additional year in August 2023 (OGoRNM No. 183/23) (Table 1.1, Fig. 2.8). Eurocode 8 defines the format and the procedure for the definition of seismic design effects upon structures. In that regard, in 2020, the seismic zoning maps published by the Institute of Earthquake engineering and Engineering Seismology in Skopje, developed by Milutinovic et al. (2016), were officially adopted under National Annex MKC EN 1998-1/HA:2020 (Fig. 2.8). The seismotectonic model backing the elaboration of listed maps in all details is presented in Dumurdjanov et al. (2020). Developed maps comply with EN 1998-1:2004 requirements in terms of (1) reference peak ground acceleration on type A ground (a_{gR}) and (2) reference return periods (T_{DLR} and T_{NCR}). Subscripts DLR and NCR stand for seismic action corresponding to damage limitation (T_{DLR}=95 years/ P_{DLR}=10% in 10 years) and no-collapse (T_{NCR}=475 years/T_{NCR}=10% in 50 years) requirements, respectively.

The 95-year return period (DLR) map divides RNM into three a_{gR} seismic zones (Z-1 to Z-3), ranging from a_{gR} = 0.05g (zone Z-1) to a_{gR} = 0.15g (zone Z-3), g=9.81 m/s^2. The solid black lines are acceleration isolines, i.e., the results from rigorous seismic hazard assessment mapped on equidistant acceleration step of 0.05g. Regarding this map, from the total country area, 5.91% is related to a_{gR}=0.15g, 50.73% to a_{gR}=0.10g, and 43.36% is related to a_{gR}=0.05g. The 475-year return period (NCR) map divides RNM into five a_{gR} seismic zones (Z-1 to Z-5). a_{gR} ranges from a_{gR} = 0.1g (zone Z-1) to a_{gR} = 0.30g (zone Z-5). NCR SZM2020 assigns 7.31% of country area to a_{gR}=0.30 g, 14.07% to a_{gR}=0.2 5g, 30.57% to a_{gR}=0.20g, 40.85% to a_{gR}=0.15g, and 7.20% to a_{gR}=0.10g. DLR and NCR 2020 SZM's and reference ground accelerations for type A ground conditions for major RNM towns, the City of Skopje included, are presented in Fig. 2.8.

The novelty of DLR/NCR 2020 zoning—following the cadastral municipality borders—supports the administrative inclusion of location-specific seismic design parameters into the official administrative document "urbanistic conditions for construction," issued by municipal urbanistic offices. According to the law for construction, it is a document indispensable for the beginning of any design/

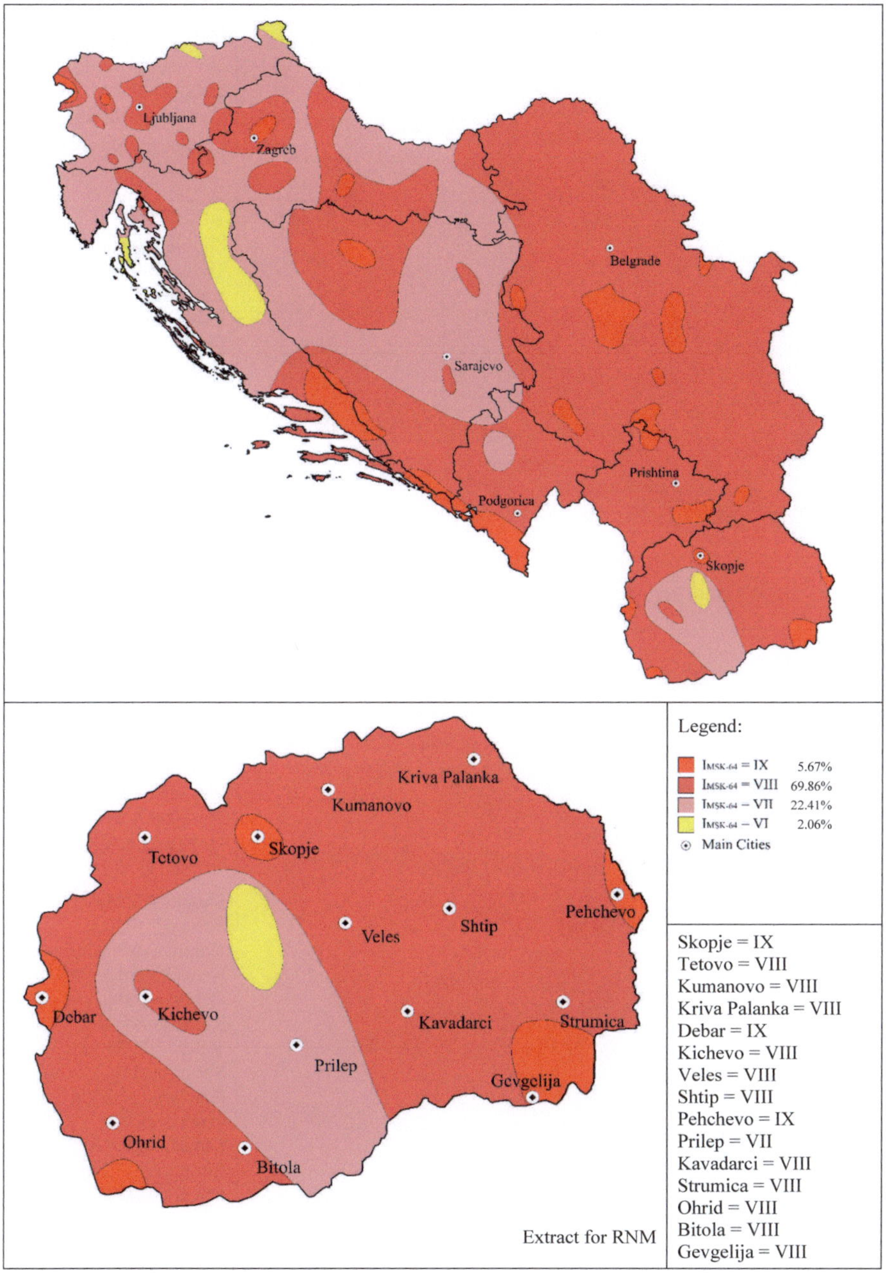

Fig. 2.7 500 years' return period seismic zoning map of SFRY (SZM1990) elaborated in 1987 and adopted in 1990 to accompany JUS31/81 seismic design code

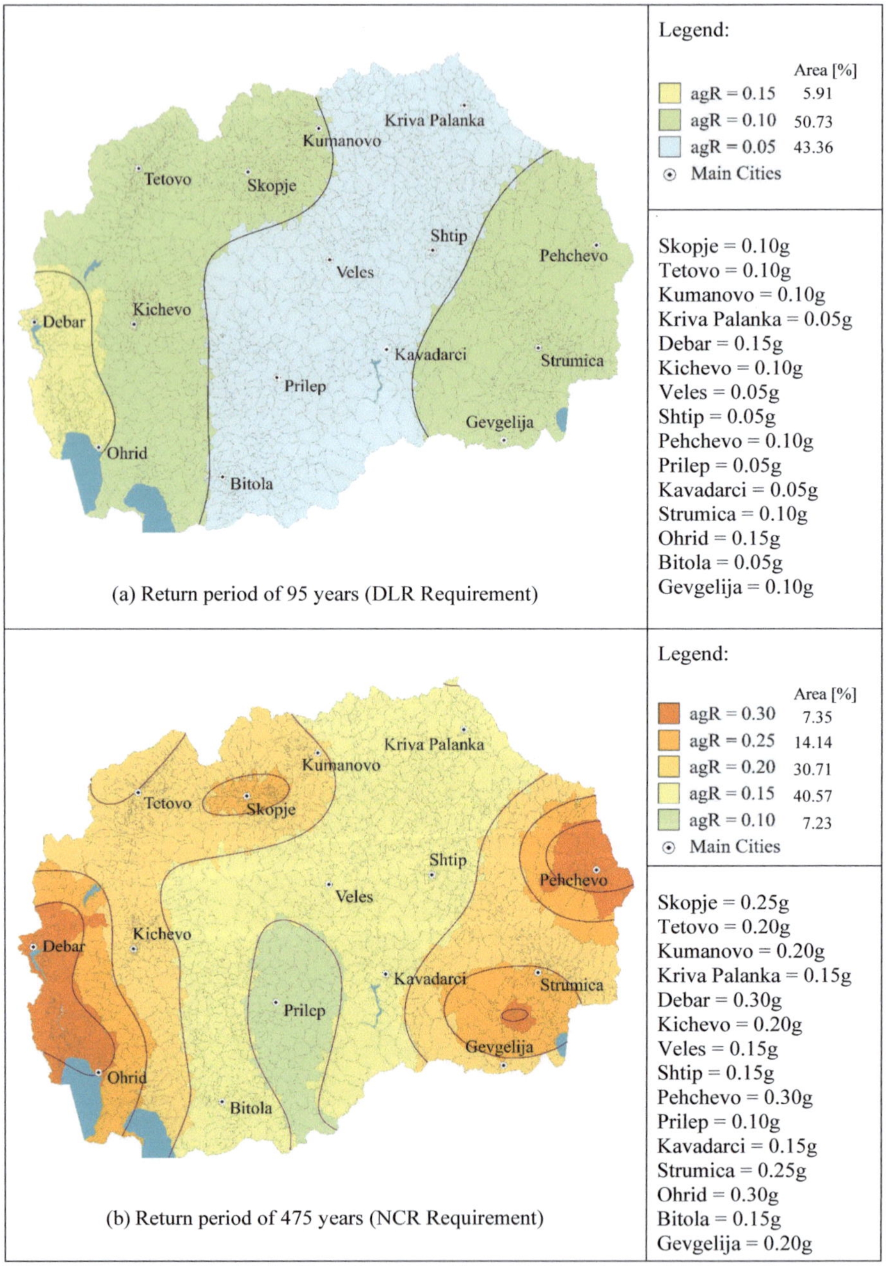

(a) Return period of 95 years (DLR Requirement)

(b) Return period of 475 years (NCR Requirement)

Fig. 2.8 DLR (a) and NCR (b) seismic zoning maps of Republic Macedonia with elements of seismic hazard (SZM2020) elaborated in 2016 and adopted in 2020 by MKC EN 1998-1/NA:2020 to accompany MKS EN 1998-1:2012 Eurocode 8

construction activities and obligatory for obtaining the construction license. Thin lines in Fig. 11 are borders of so-called "cadastral municipalities" that divide RNM into 1776 land units that comply with historic and current land book records. Such zoning by "cadastral municipalities" assures that designer, reviewer, and super reviewer (present administrative system) work under the harmonized criteria.

The 2020 DLR and NCR probabilistic seismic hazard assessments are based on two different seismotectonic models, the spatially zoned (Dumurdjanov et al. 2020) and spatially smoothed seismicity. Four ground motion prediction models and different source parameters were varied and combined, implementing a logic tree of 40 branches (Milutinovic et al. 2016).

According to the authors, applied methodological approach and assumptions made resulted in:

- Overcoming subjectivism explicitly embedded in models of spatially zoned seismicity.
- Incorporating the latest knowledge on the accumulation and kinematics of crustal deformations and active tectonics in the Southern Balkans region obtained by GPS measurements (Dumurdjanov et al. 2004, 2005, 2016; Burchfiel et al. 2006; Kotzev et al. 2008).
- Incorporating aleatory and epistemic uncertainties in the seismic hazard analysis.
- Overcoming the non-complementarity of the existing methods and techniques for seismic hazard assessment, basically compatible with the JUS31/81 code requirements, to generate products in conformity with MKC EN 1998 1: 2004: Eurocode 8.

References

Burchfiel BC, Todosov A, King RW, Kotzev V, Dumurdjanov N, Serafimovski T, Nurce B (2006) GPS results for Macedonia and its importance for the tectonics of the Southern Balkan extensional regime. Tectonophysics 413:239–248

Cubrilovic P, Palavestric L, Nikolic T (1967) Engineering Geology Map of SFR Yugoslavia (1:500,000). Federal Geological Institute, Belgrade (In Serbian)

Dumurdjanov N, Serafimovski T, Burchfiel CB (2004) Evolution of the Neogene-Pleistocene Basins of Macedonia. Geological Society of America, DMC001, pp 1–XX

Dumurdjanov N, Serafimovski T, Burchfiel CB (2005) Cenozoic tectonics of Macedonia and its relation to the South Balkan extensional regime. Geosphere 1:1–22. https://doi.org/10.1130/GES00006.1

Dumurdjanov N, Milutinovic Z, Salic R (2016) Seismotectonic zones and seismic hazard of Republic of Macedonia, Third Congress of Geologists of Republic of Macedonia, 30.09–02.10.2016, Struga, Macedonia (In Macedonian)

Dumurdjanov N, Milutinovic Z, Salic R (2020) Seismotectonic model backing the PSHA and seismic zoning of Republic of Macedonia for National Annex to MKS EN 1998-1:2012 Eurocode 8. J Seismol. https://doi.org/10.1007/s10950-020-09912-9

EN 1998-1 (2004) Eurocode 8: Design of structures for earthquake resistance – Part 1: General rules, seismic actions and rules for buildings [Authority: The European Union Per Regulation 305/2011, Directive 98/34/EC, Directive 2004/18/EC]

Kotzev V, King RW, Burchfiel BC, Todosov A, Nurce B, Nakov R (2008) Crustal Motion and Strain Accumulation in the South Balkan Region Inferred from GPS Measurements. In Husebye, E., ed., Earthquake Monitoring and Seismic Hazard Mitigation in Balkan Countries: Proceedings of the NATO Advanced Research Workshop on Earthquake Monitoring and Seismic Hazard Mitigation in Balkan Countries, Borovetz, Bulgaria, 11–18 September 2005: NATO Science Series IV: Earth and Environmental Sciences. Volume 81, 19–43

Milutinovic Z, Salic R, Tomic D (2016) Seismic zoning map of the Republic of Macedonia according to the requirements of MKC EN 1998-1:2004—Eurocode 8 (map reading guide), Report IZIIS 2016-26, August 2016 (in Macedonian)

Milutinovic Z, Salic Makreska R, Tomic D, Trajcevski J (2022) Genealogy of development and codification of Yugoslavian earthquake resistant design. Bull Earthquake Eng 20:7513–7559. https://doi.org/10.1007/s10518-022-01493-x

SZM1948: Division of the territory of Federal People's Republic of Yugoslavia into seismological zones, OGoFNRY No. 61/48 of July 17, 1948 (In Serbian).

SZM1964: S.F.R. Yugoslavia Seismological Map, OGoSFRY No. 39/64 of September 30, 1964.

SZM1979: Seismic Zoning Map of S.R. Macedonia, OGoSRM No. 2/79 of January 31, 1979.

SZM1982: Provisional Seismological Map of SFRY, OGoSFRY No. 49/82 of August 13, 1982

SZM1990: Socialist Federal Republic of Yugoslavia, Seismological Maps for Return periods of 50, 100, 200, 500, 1000 and 10000 years, OGoSFRY No. 52/90 of September 7, 1990

SZM2020: Republic of Macedonia, Seismic Zoning Map with Elements of Seismic Hazard for Return Periods of 95 and 475 years, MKC EN 1998-1/HA:2020

Chapter 3
Research Studies

Abstract This chapter reviews national-scale probabilistic seismic hazard assessment (PSHA) studies that define seismic hazard for the territory of North Macedonia. It first discusses seminal domestic works, starting with Mihailov in (Towards stochastic modelling of seismicity, Doctoral Thesis, Zagreb, 1978), who developed six stochastic models of seismicity combining aerial and line sources and produced PSHA maps for return periods of 50–1000 years. Subsequent studies by Milutinovic et al. in (Spatial plan of Republic of Macedonia, conditions for occurrence and protection from seismic disasters, IZIIS Report 98-29, 1998) and Jordanovski et al. (1998) generated national hazard maps for the Spatial Plan of the Republic of Macedonia based on macroseismic catalogues and MSK-64 intensities. Salic in (Advanced approach to seismic hazard assessment in the Republic of Macedonia, Institute of Earthquake Engineering and Engineering Seismology (UKIM-IZIIS), Ss. Cyril and Methodius University in Skopje, Doctoral dissertation. Skopje, R. Macedonia, March 2015) further advanced PSHA for RNM by integrating geological, neotectonic, seismological, strong-motion and GPS data, introducing a qualitatively new approach to modeling spatial seismicity patterns. The chapter also summarizes foreign studies (AIR Worldwide 2013; Lee and Trifunac in Soil Dyn Earthq Eng 100:504–517, 2017) that provide independent PSHA maps for RNM using global or regional GMPEs and high-resolution datasets. Collectively, these studies provide a rich but heterogeneous set of national hazard models that are essential for understanding methodological uncertainties and for benchmarking official zoning maps.

Keywords Probabilistic seismic hazard assessment · Stochastic seismicity models · Macroseismic data · Strong-motion records · National hazard maps · Ground-motion prediction models · Republic of North Macedonia

Several research studies have been published to define the seismic hazard at the national level for various purposes. Additionally, many other seismic hazard studies have been conducted to define the seismic hazard along specific corridors (such as gas and oil pipelines), capital facilities, and sites for all buildings that, according to JUS39/64 and JUS31/81, were required to be designed with an importance factor of

R. Salic Makreska and Z. Milutinovic, *Seismic Hazard Zonation for the Republic of North Macedonia*, SpringerBriefs in Geography,
https://doi.org/10.1007/978-3-032-13945-0_3

1.5. The Institute of Earthquake Engineering and Engineering Seismology (IZIIS) has predominantly performed seismic hazard analysis for citing documentation since its establishment in 1965 (including works by Mihailov, Stojkovic, Dimiskovska, Dojcinovski, Stamatovska, Milutinovic, Salic, and others). This section presents and discusses published seismic hazard mappings conducted at the national level only focusing on PSHA studies, although some efforts have been also made in Neo-deterministic Seismic Hazard Analysis (NDSHA) (e.g., Hasi 2014; Milkova et al. 2019).

3.1 Research Studies by Domestic Authors

3.1.1 Mihailov (1978)

The research presented in Mihailov (1978) is part of his doctoral thesis entitled "Contribution to Stochastic Seismic Modelling," realized in the University of Zagreb, Faculty of Natural Sciences and Mathematics, Croatia, under supervision of Prof. Dr. Dragutin Skoko and Prof. Berislav Makjanic. This research is a pioneer work not only as of the first of its kind since the establishment of IZIIS but also on the national and SFRY level considering the probabilistic seismic hazard analysis.

By a combination of parameters of seismicity, six stochastic models of seismicity for RNM were defined and analyzed in all detail, viz., cumulative distribution functions of the maximum probabilistic ground acceleration defined and applied for the elaboration of seismic hazard maps, a probability distribution for maximum acceleration occurrence for a return period of 50, 100, 500, and 1000 years. Adopted six seismotectonic models are dominantly different combinations of aerial and line sources with different treatment of recurrent parameters.

The final maps and results are presented in Fig. 3.1 and Table 3.1. Spatial distribution of the contoured results for all adopted models shows acceleration values varying from 0.10 to 0.45 g for RP50, 0.10 to 0.65 g for RP100, 0.15 to 0.90 g for RP500, and 0.20 to 0.90 g for RP1000.

3.1.2 Milutinovic et al. (1998) and Jordanovski et al. (1998)

Sets of seismic hazard zoning maps presented in Figs. 3.2 and 3.3 are developed under the study entitled "Conditions for the occurrence of seismic and technical-technological catastrophes" (Milutinovic et al. 1998), based on the Agreement between the Ministry of Urbanism, Construction, and Environmental Protection and the Institute of Earthquake Engineering and Engineering Seismology (IZIIS). The segment "Conditions for Occurrence and Protection from Seismic Disasters (*Report IZIIS 98-29*)" is elaborated in cooperation with Seismological Observatory, Faculty

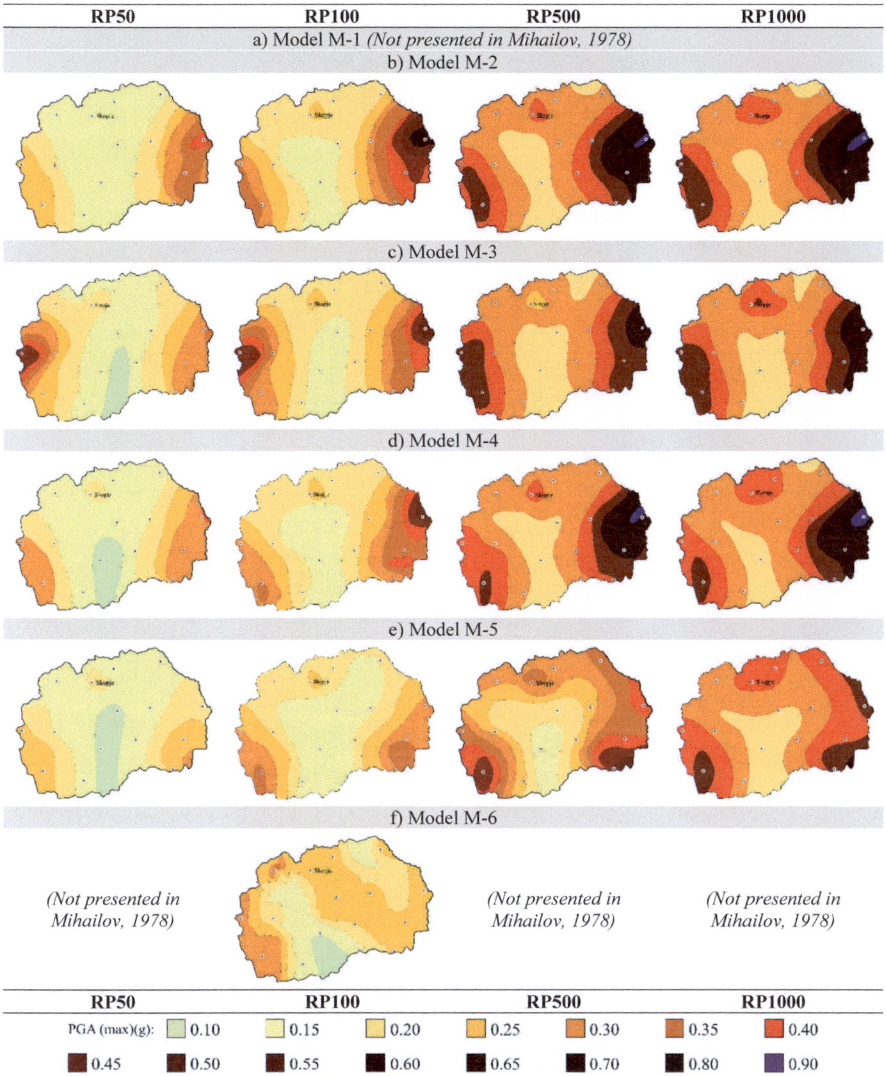

Fig. 3.1 Seismic hazard maps according to Mihailov (1978)

of Natural Sciences and Mathematics, Ss. Cyril and Methodius University in Skopje. Both sets of maps were developed for the 2000 revision of the Spatial Plan of Republic of Macedonia.

Jordanovski et al. (1998), compiled all macroseismic data available for the territory of RNM and a period 1900–1996 and, based on the Extreme Value Theory, generated MSK-64 seismic hazard maps for return periods of 100, 200, 500, 1000, and 10,000

Table 3.1 PGA (g) results for selected cities in RNM according to Mihailov (1978)

Model	Return period	Town/city						
		Skopje	Tetovo	Kumanovo	Kriva Palanka	Debar	Kichevo	Veles
1	50–1000	*(Not presented in* Mihailov 1978*)*						
2	50	0.10–0.15	0.10–0.15	0.10–0.15	0.10–0.15	0.20–0.25	0.15–0.20	0.10–0.15
	100	0.20–0.25	0.15–0.20	0.15–0.20	0.15–0.20	0.25–0.30	0.15–0.20	0.15–0.20
	500	0.30–0.40	0.20–0.30	0.20–0.30	0.20–0.30	0.30–0.40	0.20–0.30	0.20–0.30
	1000	0.30–0.40	0.20–0.30	0.30–0.40	0.20–0.30	0.40–0.50	0.20–0.30	0.20–0.30
3	50	0.15–0.20	0.15–0.20	0.10–0.15	0.15–0.20	0.40–0.45	0.25–0.30	0.10–0.15
	100	0.20–0.25	0.15–0.20	0.15–0.20	0.15–0.20	0.40–0.45	0.25–0.30	0.10–0.15
	500	0.20–0.25	0.25–0.30	0.25–0.30	0.25–0.30	0.40–0.50	0.30–0.40	0.15–0.20
	1000	0.30–0.40	0.20–0.30	0.20–0.30	0.20–0.30	0.40–0.50	0.30–0.40	0.10–0.20
4	50	0.15–0.20	0.10–0.15	0.10–0.15	0.10–0.15	0.25–0.30	0.15–0.20	0.10–0.15
	100	0.20–0.25	0.15–0.20	0.15–0.20	0.15–0.20	0.25–0.30	0.15–0.20	0.10–0.15
	500	0.30–0.40	0.20–0.30	0.20–0.30	0.20–0.30	0.30–0.40	0.20–0.30	0.10–0.20
	1000	0.30–0.40	0.20–0.30	0.30–0.40	0.20–0.30	0.30–0.40	0.20–0.30	0.20–0.30
5	50	0.15–0.20	0.10–0.15	0.10–0.15	0.10–0.15	0.20–0.25	0.10–0.15	0.05–0.10
	100	0.20–0.25	0.15–0.20	0.15–0.20	0.15–0.20	0.25–0.30	0.15–0.20	0.10–0.15
	500	0.30–0.35	0.25–0.30	0.25–0.30	0.30–0.35	0.30–0.35	0.20–0.25	0.15–0.20
	1000	0.30–0.40	0.20–0.30	0.30–0.40	0.30–0.40	0.30–0.40	0.20–0.30	0.10–0.20
6	50	*(Not presented in* Mihailov 1978*)*						
	100	0.20–0.25	0.25–0.30	0.20–0.25	0.20–0.25	0.25–0.30	0.15–0.20	0.20–0.25
	500	*(Not presented in* Mihailov 1978*)*						
	1000							

(continued)

Table 3.1 (continued)

Model	Return period	Town/city							
		Shtip	Pehchevo	Prilep	Kavadarci	Strumica	Ohrid	Bitola	Gevgelija
1	50–1000	*(Not presented in* Mihailov 1978*)*							
2	50	0.15–0.20	0.35–0.40	0.10–0.15	0.15–0.20	0.30–0.35	0.20–0.25	0.10–0.15	0.20–0.25
	100	0.25–0.30	0.60–0.65	0.10–0.15	0.20–0.25	0.35–0.40	0.25–0.30	0.15–0.20	0.30–0.35
	500	0.30–0.40	0.80–0.90	0.10–0.20	0.30–0.40	0.60–0.70	0.40–0.50	0.20–0.30	0.40–0.50
	1000	0.40–0.50	0.80–0.90	0.10–0.20	0.30–0.40	0.40–0.50	0.40–0.50	0.20–0.30	0.40–0.50
3	50	0.10–0.15	0.25–0.30	0.10–0.15	0.10–0.15	0.25–0.30	0.20–0.25	0.10–0.15	0.20–0.25
	100	0.20–0.25	0.45–0.50	0.10–0.15	0.15–0.20	0.30–0.35	0.25–0.30	0.15–0.20	0.25–0.30
	500	0.25–0.30	0.50–0.60	0.15–0.20	0.25–0.30	0.40–0.50	0.40–0.50	0.25–0.30	0.40–0.50
	1000	0.20–0.30	0.65–0.70	0.10–0.20	0.20–0.30	0.40–0.50	0.40–0.50	0.20–0.30	0.40–0.50
4	50	0.15–0.20	0.25–0.30	0.05–0.10	0.10–0.15	0.25–0.30	0.20–0.30	0.10–0.15	0.25–0.30
	100	0.20–0.25	0.45–0.50	0.10–0.15	0.15–0.20	0.30–0.35	0.30–0.35	0.15–0.20	0.25–0.30
	500	0.20–0.30	0.70–0.80	0.10–0.20	0.20–0.30	0.50–0.60	0.40–0.50	0.20–0.30	0.30–0.40
	1000	0.30–0.40	0.80–0.90	0.10–0.20	0.30–0.40	0.60–0.70	0.40–0.50	0.20–0.30	0.40–0.50
5	50	0.10–0.15	0.15–0.20	0.05–0.10	0.10–0.15	0.20–0.25	0.20–0.25	0.10–0.15	0.20–0.25
	100	0.10–0.15	0.25–0.30	0.10–0.15	0.15–0.20	0.25–0.30	0.30–0.35	0.10–0.15	0.25–0.30
	500	0.20–0.25	0.35–0.40	0.10–0.15	0.20–0.25	0.35–0.40	0.40–0.45	0.20–0.25	0.40–0.45
	1000	0.20–0.30	0.40–0.50	0.10–0.20	0.20–0.30	0.30–0.40	0.40–0.50	0.20–0.30	0.40–0.50
6	50	*(Not presented in* Mihailov 1978*)*							
	100	0.20–0.25	0.20–0.25	0.10–0.15	0.20–0.25	0.20–0.25	0.25–0.30	0.10–0.15	0.20–0.25
	500	*(Not presented in* Mihailov 1978*)*							
	1000								

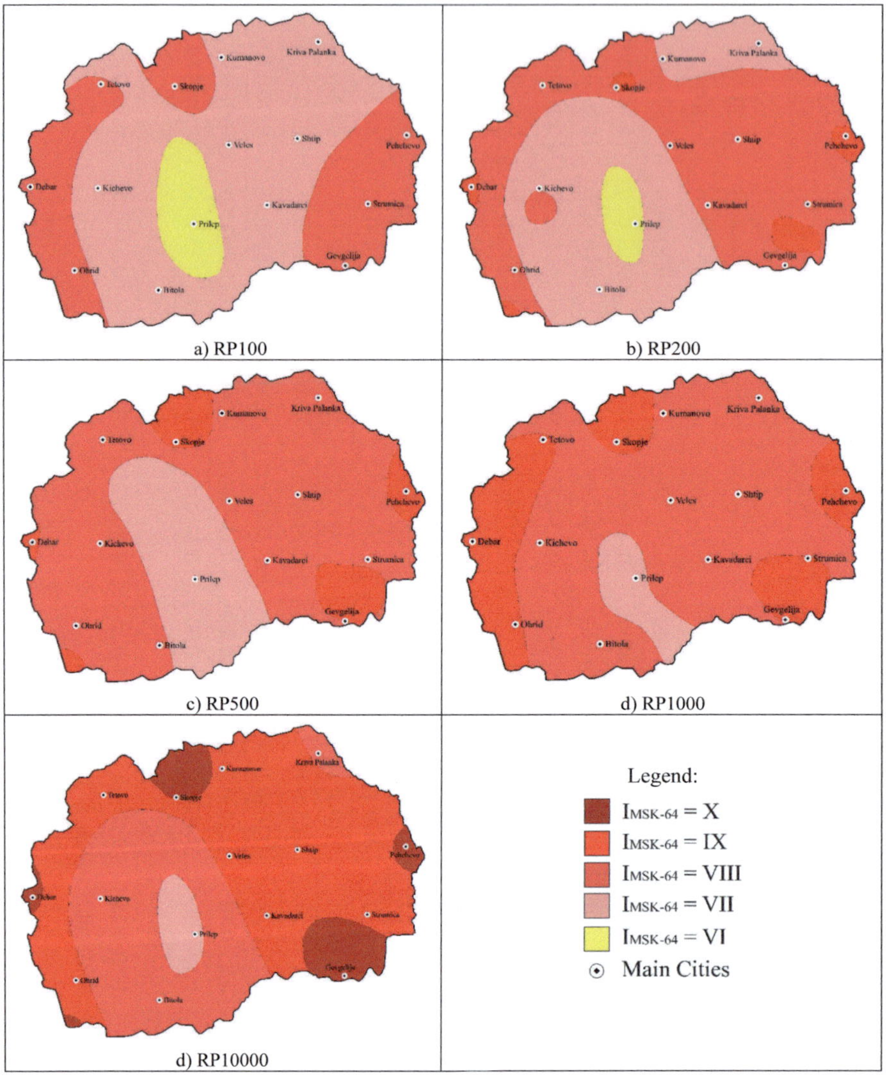

Fig. 3.2 Seismic hazard maps according to spatial plan of the Republic of Macedonia, conditions for occurrence and protection against seismic disasters (Jordanovski et al. 1998)

years, Fig. 3.2. Presented contoured MSK-64 values range from VI to VIII for RP100, VI to IX for RP200, and VII to IX for RP500 and RP1000.

The map of maximum observed intensities is produced as an intermediate result. It indicates that in the twentieth century (1900–1996), the entire territory of RNM was exposed to minimum MSK-64 intensity of VI. More specifically, 3.0% of the country area was exposed to X MSK-64 intensity, 11.0% to IX MSK-64, 38.2% to VIII MSK-64, 45.6% to VII MSK-64, and 2.2% to VI MSK-64 intensity.

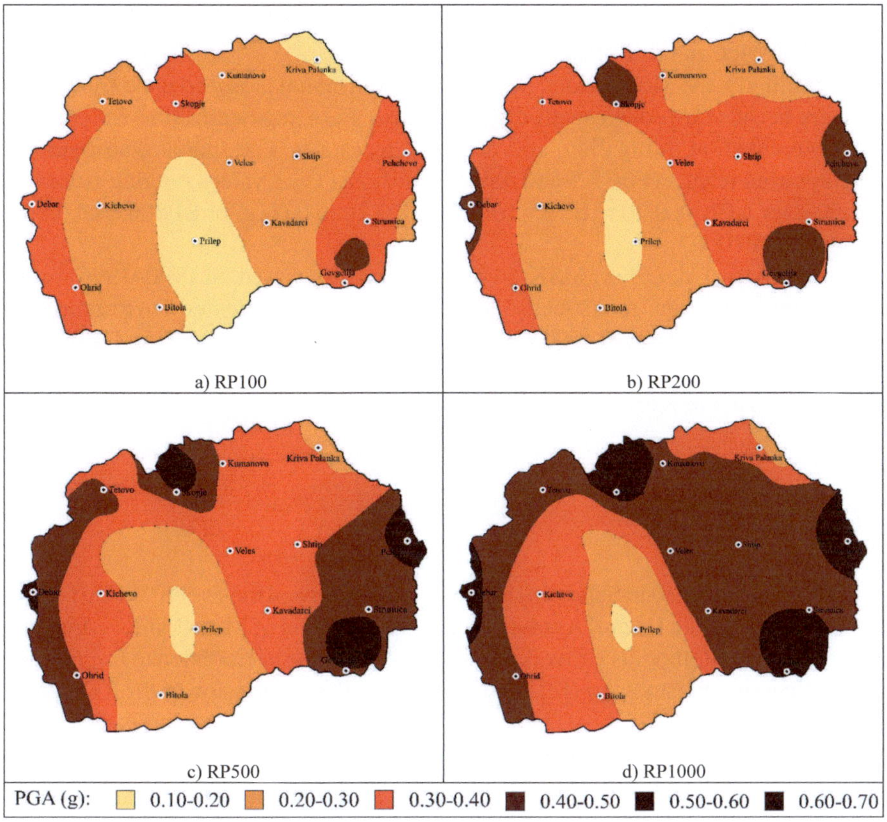

Fig. 3.3. Seismic hazard maps according to spatial plan of the Republic of Macedonia, conditions for occurrence and protection against seismic disasters (Milutinovic et al. 1998)

Peak ground acceleration (PGA) maps developed by Milutinovic et al. (1998), are obtained by converting each RP-related expected maximum MSK-64 intensity maps into the corresponding maximum expected PGA maps, Fig. 3.3. For conversion are tested five PGA—MSK-64 relations (Gutenberg and Richter 1942; Herschberger 1956; Ambraseys 1974; Trifunac and Brady 1975; Murphy and O'Brien 1977; McCann and Shah 1979). In the MSK-64 intensity domain V–VIII, all conversion relations produce comparable PGA's.

However, their tail behavior over the intensity range VII–X completely differs. Ambraseys (1974) and Herschberger (1956) provide up to 6 times higher PGA values than McCann and Shah (1979) conversion. Relative to McCann and Shah (1979), Trifunac and Brady (1975) conversion up to MSK-64 IX intensity provides lower, but for MSK-64 = X, provides for about 30% higher PGA. Gutenberg and Richter (1942) and Murphy and O'Brien (1977) compared with McCann and Shah (1979) for MSK-64 = X provide ~ 10% and 30% lower PGAs, respectively. They overlap over the MSK-64 VIII to IX intensity range, and both provide ~ 50% lower PGA values than McCann and Shah (1979).

Accordingly, the McCann and Shah (1979) conversion is agreed upon by regional planners because it provides sufficiently distinct PGA regionalization for planning various prevention, mitigation, and response measures. The spatial distribution of the contoured results (Fig. 3.3 and Table 3.3) shows PGA values ranging from 0.20 to 0.50 g for RP100, 0.20 to 0.50 g for RP200, 0.20 to 0.60 g for RP500, and 0.20 to 0.70 g for RP1000.

Jordanovski et al. (1998) MSK-64 intensity zoning, based on 1900–1996 macroseismic intensity data, realistically quantify Macedonian seismic environment for return periods of 100, 200, 500, 1000, and 10,000 years, Fig. 3.2 and Table 3.2. Milutinovic et al. (1998) PGA maps (Fig. 3.3 and Table 3.3) should be considered as maps targeted to meet planners' needs.

3.1.3 Salic (2015)

The research presented in Salic (2015) is part of her doctoral thesis entitled "Advanced Approach to Seismic Hazard Assessment for the Republic of Macedonia," completed at Ss. Cyril and Methodius University in Skopje, Institute of Earthquake Engineering and Engineering Seismology (IZIIS), RNM, under the supervision of Prof. Dr. Zoran Milutinovic.

Salic's work is a comprehensive upgrade of the probabilistic seismic hazard assessment (PSHA) for RNM. An integrated analysis of data from geological, tectonic, neo-tectonic, and seismological investigations was combined with existing regional seismic hazard studies, databases of national and regional strong motion records, and GPS measurements. This approach introduced a qualitatively new method for modeling the spatial distribution of seismicity.

Salic's approach emphasizes the importance of multidisciplinary data integration, which enhances the robustness of seismic hazard assessments. Innovative techniques used have set new standards in the field, contributing to more accurate and reliable seismic hazard predictions. Furthermore, Salic's research underscores the critical role of advanced modeling techniques in understanding seismic hazard, providing a foundation for future studies in this domain.

Three different analytical models for seismic hazard assessment were developed: the gridded seismicity model (referred to as M-1), the zoned seismicity model with source delineation using the distribution of historical seismicity (M-2), and the analysis of seismotectonic features (M-3). The recurrence parameters were defined using different approaches. Four ground motion prediction equations (GMPEs) were implemented as models suitable for the seismotectonic settings in Southern Europe. The analyses were conducted for type A ground conditions and four distinct exceedance probability levels: 10% in 10 years (RP95), 10% in 50 years (RP475), 2% in 50 years (RP2475), and 1% in 100 years (RP10,000).

Table 3.2 $I_{\text{MSK-64}}$ values for selected cities in RNM according to Jordanovski et al. (1998)

Return period	Town/city														
	Skopje	Tetovo	Kumanovo	Kriva Palanka	Debar	Kichevo	Veles	Shtip	Pehchevo	Prilep	Kavadarci	Strumica	Ohrid	Bitola	Gevgelija
100	VIII	VIII	VII	VII	VIII	VII	VII	VII	VIII	VI	VII	VIII	VIII	VII	VIII
200	IX	VIII	VII	VII	IX	VII	VIII	VIII	IX	VI	VIII	VIII	VIII	VII	VIII
500	IX	VIII	VIII	VIII	IX	VIII	VIII	VIII	IX	VII	VIII	VIII	VIII	VIII	IX
1000	IX	IX	VIII	VIII	IX	VIII	VIII	VIII	IX	VII	VIII	IX	IX	VIII	IX
10,000	X	IX	IX	VIII	X	VIII	IX	IX	X	VII	IX	IX	IX	VIII	X

Table 3.3 PGA (g) values for selected cities in RNM according to Milutinovic et al. (1998)

Return period	Town/city						
	Skopje	Tetovo	Kumanovo	Kriva Palanka	Debar	Kichevo	Veles
100	0.30–0.40	0.20–0.30	0.20–0.30	0.10–0.20	0.30–0.40	0.20–0.30	0.20–0.30
200	0.40–0.50	0.30–0.40	0.20–0.30	0.20–0.30	0.40–0.50	0.20–0.30	0.30–0.40
500	0.40–0.50	0.40–0.50	0.30–0.40	0.20–0.30	0.50–0.60	0.30–0.40	0.30–0.40
1000	0.50–0.60	0.40–0.50	0.40–0.50	0.30–0.40	0.60–0.70	0.30–0.40	0.40–0.50

Return period	Town/city							
	Shtip	Pehchevo	Prilep	Kavadarci	Strumica	Ohrid	Bitola	Gevgelija
100	0.20–0.30	0.30–0.40	0.10–0.20	0.20–0.30	0.30–0.40	0.30–0.40	0.20–0.30	0.30–0.40
200	0.30–0.40	0.40–0.50	0.10–0.20	0.30–0.40	0.30–0.40	0.30–0.40	0.20–0.30	0.40–0.50
500	0.30–0.40	0.50–0.60	0.20–0.30	0.30–0.40	0.40–0.50	0.40–0.50	0.20–0.30	0.40–0.50
1000	0.40–0.50	0.60–0.70	0.20–0.30	0.40–0.50	0.50–0.60	0.40–0.50	0.30–0.40	0.50–0.60

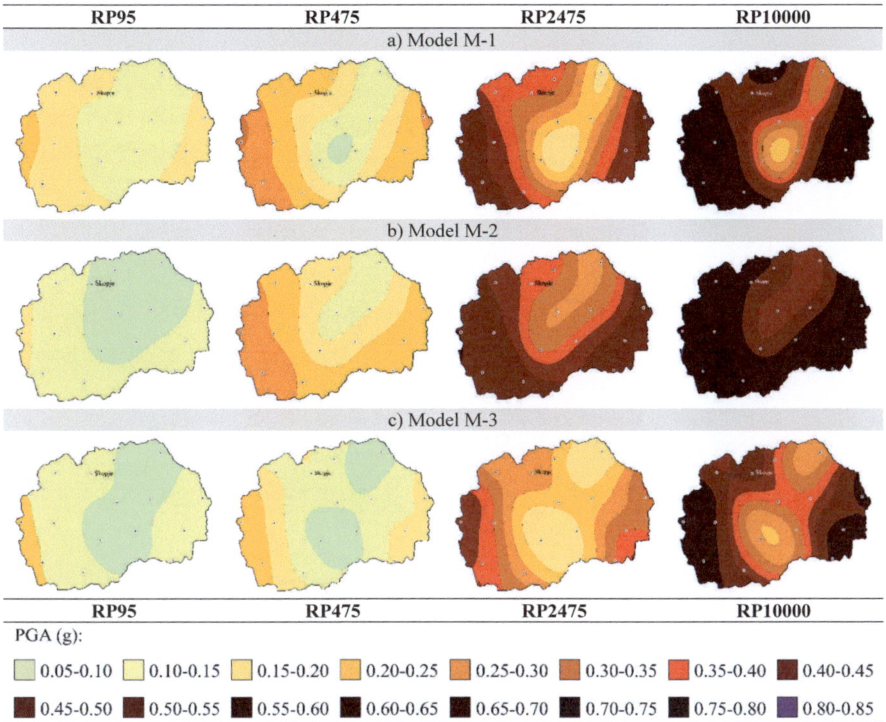

Fig. 3.4 Seismic hazard maps according to Salic (2015)

The probabilistic seismic hazard maps and results showing the median a_{gR} values for defined return periods and soil type A (V_s30) = 800 m/s) are presented in Fig. 3.4. The ground type A, a_{gR}, values estimated for major RNM towns are presented in Table 3.4. Spatial distribution of the contoured results shows values ranging from 0.10 to 0.25 g for RP95, 0.10 to 0.35 g for RP475, 0.20 to 0.60 g for RP2475, and 0.25 to 0.85 g for RP10,000 for all presented models.

Salic's (2015) work served as the basis for developing the 2020 DLR and NCR probabilistic seismic hazard assessments adopted by MKC EN 1998-1/HA (2020) as the National Annex to MKC EN 1998-1.

3.2 Research Studies by Foreign Authors

An extensive internet survey revealed only two recent studies conducted by foreign authors and research groups that encounter the whole territory of RNM, the AIR Worldwide (2013) and Lee and Trifunac (2017).

Table 3.4 a_{gR} (g) values for selected cities in RNM according to Salic (2015)

Model	Return period	Town/city														
		Skopje	Tetovo	Kumanovo	Kriva Palanka	Debar	Kichevo	Veles	Shtip	Pehchevo	Prilep	Kavadarci	Strumica	Ohrid	Bitola	Gevgelija
1	95	0.11	0.12	0.11	0.05	0.17	0.13	0.06	0.06	0.12	0.06	0.05	0.11	0.15	0.11	0.11
	475	0.21	0.22	0.21	0.10	0.32	0.24	0.12	0.11	0.26	0.10	0.09	0.22	0.27	0.22	0.22
	2475	0.37	0.38	0.38	0.19	0.53	0.42	0.22	0.21	0.50	0.17	0.17	0.43	0.48	0.40	0.41
	10,000	0.54	0.56	0.58	0.30	0.78	0.64	0.34	0.34	0.81	0.25	0.26	0.70	0.71	0.61	0.64
2	95	0.10	0.11	0.10	0.06	0.16	0.12	0.07	0.07	0.11	0.08	0.08	0.11	0.15	0.12	0.11
	475	0.19	0.21	0.19	0.12	0.29	0.24	0.13	0.14	0.22	0.15	0.16	0.22	0.28	0.23	0.21
	2475	0.34	0.35	0.34	0.23	0.49	0.41	0.24	0.25	0.43	0.26	0.29	0.43	0.48	0.41	0.40
	10,000	0.52	0.53	0.52	0.35	0.74	0.61	0.36	0.38	0.71	0.39	0.46	0.71	0.72	0.62	0.65
3	95	0.06	0.06	0.05	0.03	0.10	0.08	0.04	0.04	0.07	0.04	0.04	0.06	0.10	0.06	0.06
	475	0.16	0.14	0.11	0.07	0.22	0.17	0.10	0.10	0.17	0.08	0.09	0.17	0.20	0.13	0.13
	2475	0.30	0.27	0.22	0.14	0.42	0.33	0.21	0.21	0.35	0.15	0.17	0.35	0.38	0.26	0.28
	10,000	0.48	0.44	0.36	0.24	0.68	0.53	0.35	0.35	0.58	0.23	0.28	0.60	0.59	0.42	0.47

3.2.1 AIR Worldwide (2013)

As part of the Flood and Earthquake Risk Assessment for Albania, Serbia, and Macedonia in support of the Europa Reinsurance (EuropaRe) Facility Ltd. funded by the World Bank, AIR Worldwide has developed a country-specific probabilistic seismic risk assessment model for three countries in Southeast Europe: Albania, Macedonia, and Serbia. The framework used in the study is consistent with the AIR Worldwide 2011 Earthquake Model developed for the Pan-European Region. It covers 30 countries in Europe, including the most at risk, such as Turkey, Italy, Greece, Portugal, and, within Eastern Europe, Romania, Bulgaria, Hungary, and Slovenia, among others. The AIR Worldwide RP475 seismic hazard maps were presented by Galasso et al. in 2013 at the International Conference on Earthquake Engineering (SE-50EEE) held in Skopje, RNM, in May 2013.

For, by alphabetic order, Albania, Macedonia, and Serbia AIR Worldwide synthetically generated a set of potential earthquakes, with their temporal and geographical distribution, based on historical and instrumental earthquake catalogues, tectonic setting, plate motion information, and available data on crustal faults. Different sets of state-of-the-art ground motion prediction equations (GMPEs) were used to account for different types of earthquakes. The effect of site conditions on ground motion is modeled based on shear wave velocities defined from high-resolution topographic slope data. Two high-resolution seismic hazard maps are presented that, according to the authors, should benefit both public and private stakeholders via multiple applications, such as catastrophe risk insurance markets, urban development planning, community-based earthquake risk management and mitigation, disaster risk financing, and building code applications.

The probabilistic seismic hazard maps showing the median PGA values with 10% probability of exceedance in 50 years (475-year return period), computed on hard ground (rock) and soil, are presented in Fig. 3.5. Presented contoured values range from 0.10–0.35 g for both soil conditions.

3.2.2 Lee and Trifunac (2017)

Lee and Trifunac (2017) present seismic hazard maps for RNM and show how the Uniform Hazard Spectra (UHS) method eliminates the difficulties of using the fixed-shape design spectra. It also shows that the UHS method eliminates the issues that result from scaling in terms of peak accelerations—when it is necessary to include hazard contributions from large distant earthquakes—and the problem of the inability of the fixed-shape spectra to describe frequency-dependent spectral shapes for different probabilities of exceedance. The authors also discuss how far south contributions of the Vrancea earthquakes reach, as seen in the UHS of strong ground motion in RNM.

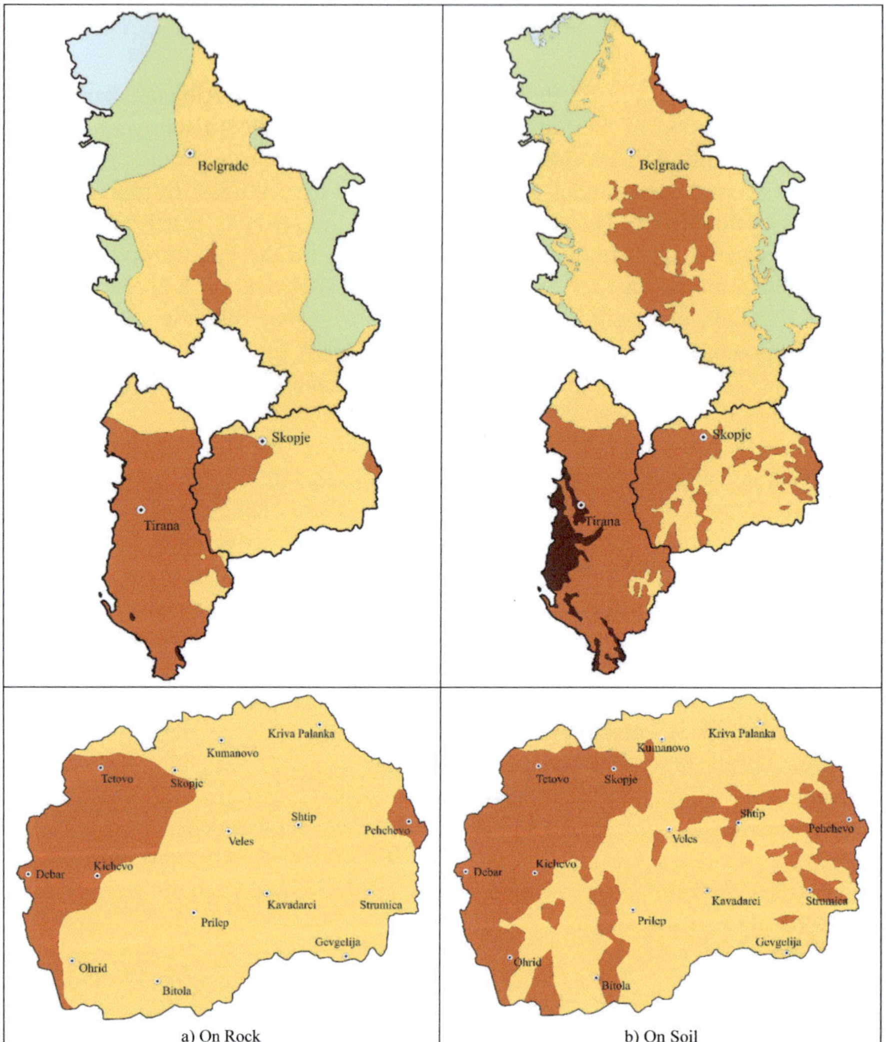

a) On Rock		b) On Soil	
Skopje = 0.10-0.20g	Pehchevo = 0.20-0.35g	Skopje = 0.20-0.35g	Pehchevo = 0.20-0.35g
Tetovo = 0.20-0.35g	Prilep = 0.10-0.20g	Tetovo = 0.20-0.35g	Prilep = 0.10-0.20g
Kumanovo = 0.10-0.20g	Kavadarci = 0.10-0.20g	Kumanovo = 0.10-0.20g	Kavadarci = 0.10-0.20g
Kriva Palanka = 0.10-0.20g	Strumica = 0.10-0.20g	Kriva Palanka = 0.10-0.20g	Strumica = 0.20-0.35g
Debar = 0.20-0.35g	Ohrid = 0.10-0.20g	Debar = 0.20-0.35g	Ohrid = 0.20-0.35g
Kichevo = 0.20-0.35g	Bitola = 0.10-0.20g	Kichevo = 0.20-0.35g	Bitola = 0.10-0.20g
Veles = 0.10-0.20g	Gevgelija = 0.10-0.20g	Veles = 0.10-0.20g	Gevgelija = 0.10-0.20g
Shtip = 0.10-0.20g		Shtip = 0.10-0.20g	

Legend: agR = 0.10-0.20g; agR = 0.20-0.35g; ⊙ Main Cities

Fig. 3.5 Seismic hazard map for return period of 475 years according to AIR Worldwide (2013)

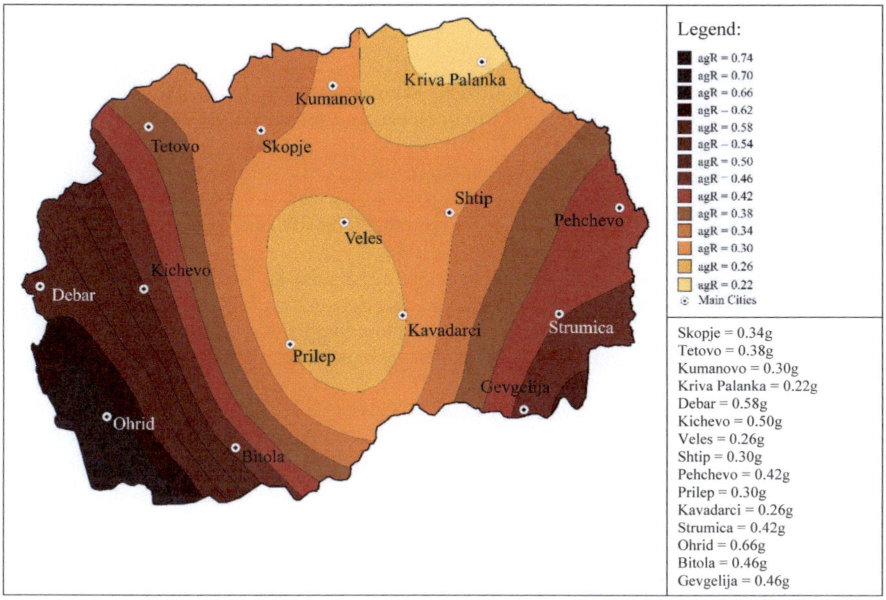

Fig. 3.6 Seismic hazard map for $T_{\mathrm{PSA}} = 0.4$ s and the return period of 475 years according to Lee and Trifunac (2017)

A final note the authors stated in the paper is that since they used relatively old empirical scaling equations for the former Yugoslavia (for a prediction of response spectra in terms of local seismicity there), and only a subset of available recordings from four Vrancea earthquakes with large magnitudes, the presented results are of a preliminary nature. When additional strong motion data become available, and when all the recording sites are assigned their geological (s) and soil (SL) site parameters, according to them, all relevant empirical models should be recalibrated, and the next generation of improved hazard maps can then be presented. Also, it is stated that the most uncertain step in the process involves the methods used to define a regional seismic activity, which leads to the Gutenberg-Richter relations and associated maximum magnitudes. This uncertainty results from the following factors: incomplete data on past seismic activity, the many changes in the definition of intensity scales that seismologists have used over time, the unpredictability involved in the conversion of these intensity scales to a common scale (selected to be M_W), imperfect declustering of the catalogue, and the very short period of modern seismological observations.

Preliminary probabilistic seismic hazard map and results for a defined return period of 475 years and PSA for $T = 0.04$ s are presented in Fig. 3.6. The spatial distribution of the contoured results shows values ranging from 0.22–0.74 g.

References

Ambraseys N (1974) The correlation of intensity with ground motion. In: Proceedings of 14th conference of European Seismological Communication, Europe, Trieste, vol 1, pp 335–341

Galasso C, Gomez I, Gupta A, Shen-Tu B (2013) Probabilistic Seismic Hazard Assessment for Albania, Macedonia and Serbia. In: International Conference on Earthquake Engineering, 29–31 May 2013, Skopje, Macedonia

Gutenberg B, Richter CF (1942) Earthquake magnitude, intensity, energy, and acceleration. Bull Seismol Soc Am 32(3):163–191. ISSN 0037-1106

Hasi N (2014) Neo deterministic approach to seismic risk analysis for the territory of Kosovo, Institute of Earthquake Engineering and Engineering Seismology (UKIM-IZIIS), Ss. Cyril and Methodius University in Skopje, Doctoral dissertation. Skopje, R. Macedonia, April 2014

Herschberger J (1956) A comparison of earthquake accelerations with intensity rating. BSSA 46(4):317–320

Jordanovski LJ, Pekevski L, Chejkovska V, Chernih D, Hristovski B, Vasilevski N (1998) Basic characteristics of seismicity in the territory of the Republic of Macedonia, Ss. Cyril and MEthedius University in Skopje, Faculty of Natural Sciences and Mathematics, Seismological Observatory, 1998 (in macedonian)

Lee VW, Trifunac MD (2017) Seismic hazard maps in Macedonia. Soil Dyn Earthq Eng 100:504–517, ISSN 0267-7261. https://doi.org/10.1016/j.soildyn.2017.06.010

McCann M, Shah H (1979) Determining strong-motion duration of earthquakes. Bull Seismol Soc Am 69(4):1253–1265. https://doi.org/10.1785/BSSA0690041253

Mihailov V (1978) Towards stochastic modelling of seismicity, Doctoral Thesis, Zagreb (in Croatian)

Milkova K, Dumova-Jovanoska E, Vaccari F, Romanelli F, Panza G (2019) Application of neo-deterministic analysis for North Macedonia. Sci J Civ Eng 8(2)

Milutinovic Z, Mihailov V, Talaganov K, Trendafiloski G, Olumceva T, Sesov V (1998) Spatial plan of Republic of Macedonia, conditions for occurrence and protection from seismic disasters, IZIIS Report 98-29 (In Macedonian)

МКС EN 1998-1/НА:2020: Еврокод 8: Проектирање конструкции отпорни на земјотрес – Дел 1: Општи правила, сеизмички дејствија и правила за згради - Национален анекс/Eurocode 8: Design of structures for earthquake resistance—part 1: General rules, seismic actions and rules for buildings—National Annex

Murphy JR, O'Brien LJ (1977) The correlation of peak ground acceleration amplitude with seismic intensity and other physical parameters. Bull Seismol Soc Am 67(3):877–915. https://doi.org/10.1785/BSSA0670030877

Salic R (2015) Advanced approach to seismic hazard assessment in the Republic of Macedonia, Institute of Earthquake Engineering and Engineering Seismology (UKIM-IZIIS), Ss. Cyril and Methodius University in Skopje, Doctoral dissertation. Skopje, R. Macedonia, March 2015 (in Macedonian)

Trifunac MD, Brady AG (1975) On the correlation of seismic intensity scales with the peaks of recorded strong ground motion. Bull Seismol Soc Am 65:139–162

Chapter 4
Research Studies Part of Regional and Global Efforts

Abstract This chapter examines regional and global seismic hazard projects that include the territory of North Macedonia and have produced harmonized hazard maps for the wider Western Balkans and Euro-Mediterranean region. It first presents the NATO Science for Peace projects BSHAP-1 and BSHAP-2 and the BALKANS-OQ initiative. BSHAP projects developed RP95 and RP475 PSHA maps using regionally consistent seismotectonic models, smoothed-gridded seismicity, and logic-tree combinations of multiple GMPEs across participating countries. The chapter then reviews global and European models—GSHAP, SESAME, and the European Seismic Hazard Models (ESHM13 and ESHM20)—highlighting their objectives, data sources, and grid-based PGA (and spectral) hazard products. For North Macedonia, these models provide independent, externally validated hazard estimates and ensure consistency with European design practice, serving as key references for the calibration and comparison of national seismic zoning.

Keywords Regional seismic hazard · Western Balkans · BSHAP-1 · BSHAP-2 · BALKANS-OQ · GSHAP · ESHM13 · ESHM20 · Harmonized hazard maps

To date, after year 2000, the seismic hazard for the territory of RNM has been assessed in several regional (Balkan/Western Balkan) and global (European/Worldwide) studies. Although conducted with regional resolution, these studies are considered referent since they were supported and performed by participation of relevant institutions, most often national authorities and universities.

4.1 Regional Level Research Studies

In the last 15 years, on a regional scale, three significant studies for defining seismic hazard were performed with the active participation of referent national institutions. These projects were of essential importance for the region, not only for their findings

R. Salic Makreska and Z. Milutinovic, *Seismic Hazard Zonation for the Republic of North Macedonia*, SpringerBriefs in Geography,
https://doi.org/10.1007/978-3-032-13945-0_4

but also for strengthening cooperation and the exchange of know-how in the domain of seismic hazard assessment between partner countries.

4.1.1 NATO SpS-983054: BSHAP-1 (2011)

The NATO SfP Project 983054 (Harmonization of Seismic Hazard Maps for the Western Balkan Countries—BSHAP-1) (Akkar et al. 2011) was proposed for financing to NATO in the framework of the Science for Peace Programme in mid-2007. On the recommendation of the Environmental Security Advisory Panel and on behalf of the NATO Science for Peace and Security Committee, the project was approved for financing in June 2007.

The project was finalized in December 2011, realized by seven prominent partner institutions:

- *Middle East Technical University, Ankara, Turkey.*
- *Montenegro Seismological Observatory, Podgorica, Montenegro.*
- *Ss. Cyril and Methodius University in Skopje, Institute of Earthquake Engineering and Engineering Seismology (IZIIS), Macedonia.*
- *Seismological Institute, later Institute of GeoSciences, Energy, Water and Environment, Tirana, Albania.*
- *Department of Geophysics, Faculty of Science, University of Zagreb, Zagreb, Croatia.*
- *Ministry of Civil Affairs, Sarajevo, Bosnia and Herzegovina.*
- *Seismological Institute of Serbia, Belgrade, Serbia.*

The presented results—the new generation of probabilistic seismic hazard maps for the entire BSHAP Region (in alphabetical order: Albania, Bosnia and Herzegovina, Croatia, N. Macedonia, Montenegro, and Serbia)—were assessed using the spatially smoothed seismicity method (Zabukovec et al. 2007). Interim seismic hazard maps were derived for four GMPE models: Bindi et al. (2009) ($w = 0.30$), Akkar and Bommer (2010) ($w = 0.30$), Boore and Atkinson (2008) ($w = 0.20$), and Cauzzi and Faccioli (2008) ($w = 0.20$). These maps were combined in a logic tree framework with the listed weight factors to capture epistemic uncertainty and generate the final DLR and NCR seismic hazard maps.

The resultant maps were expressed in terms of reference horizontal ground acceleration (a_{gR}) for return periods of 95 and 475 years, corresponding to the probability of exceedance of 10% in 10 years and 10% in 50 years, respectively. The assessment was carried out for rock conditions with an average shear wave velocity $V_S \geq 800$ m/sec in the upper 30 m of the ground section (ground type A according to Eurocode 8 ground characterization). Thus, the presented results were in full conformity with the Eurocode 8 requirements for seismic zonation and aseismic design.

Apart from the development of a new generation of probabilistic seismic hazard maps for the Western Balkans, BSHAP-1 enhanced the understanding of regional tectonics in general. This was achieved through the collection and compilation of new data, the development of innovative approaches, and the improvement of national seismic networks for all participating countries. Additionally, the project trained young researchers in seismic hazard assessment, among other benefits.

The probabilistic seismic hazard maps showing the RP95 and RP475 a_{gR} contours are presented in Figs. 4.1 and 4.2. Presented contoured a_{gR} values for RNM range from 0.04–0.14 g for RP95 and 0.06–0.24 g for RP475.

4.1.2 Balkans-OQ (2014)

Balkans-OQ Regional Seismic Hazard and Risk Modeling Initiative Project (Sevilgen et al. 2014) was founded for two years by the US Office of Foreign Disaster Assistance, supported by USGS and GEM with the active participation of an international group of scientists from eight Balkan countries (Albania, Bosnia and Herzegovina, Croatia, Kosovo, N. Macedonia, Montenegro, Serbia, and Slovenia). In fact, the BSHAP-1 group enlarged for Kosovo.

The hazard assessment in this project relies on strain rates determined from horizontal GPS velocities to infer slip rates along recognized faults. As a starting model, the SHARE (ESHM13) fault model was used, slightly adjusted by adding new faults and correcting some of the fault geometries based on the availability of new data. Hazard computations were performed using the OpenQuake (Pagani et al. 2014) software.

The main project's outcome was the RP475 seismic hazard map for type A ground conditions (Fig. 4.3). The spatial distribution of the contoured results for RNM shows values ranging from 0.10–0.60 g.

4.1.3 NATO SpS-984374: BSHAP-2 (2015)

The NATO SpS Project 984374 (Improvements in the Harmonized Seismic Hazard Maps for the Western Balkan Countries—BSHAP-2) (Gulerce et al. 2015) was financed for three years by the NATO Science for Peace and Security Programme. The project, realized by six main partner institutions, excluding the Ministry of Civil Affairs, Sarajevo, Bosnia and Herzegovina, was finalized in November 2015.

Among others, the main output of the BSHAP-2 project was the new and improved probabilistic RP95 and RP475 seismic hazard maps for the Western Balkans, obtained by implementing the smoothed-gridded seismicity approach aligned with Eurocode 8 requirements. Interim seismic hazard maps were generated for four GMPE models: Boore et al. (2014), $w = 0.30$; Chiou and Youngs (2014), $w = 0.30$; Akkar et al. (2014), $w = 0.20$; and Bindi et al. (2014), $w = 0.20$. These maps were combined in the logic tree framework with the listed weight factors (w) to capture the epistemic uncertainty and generate the final DLR and NCR seismic hazard maps.

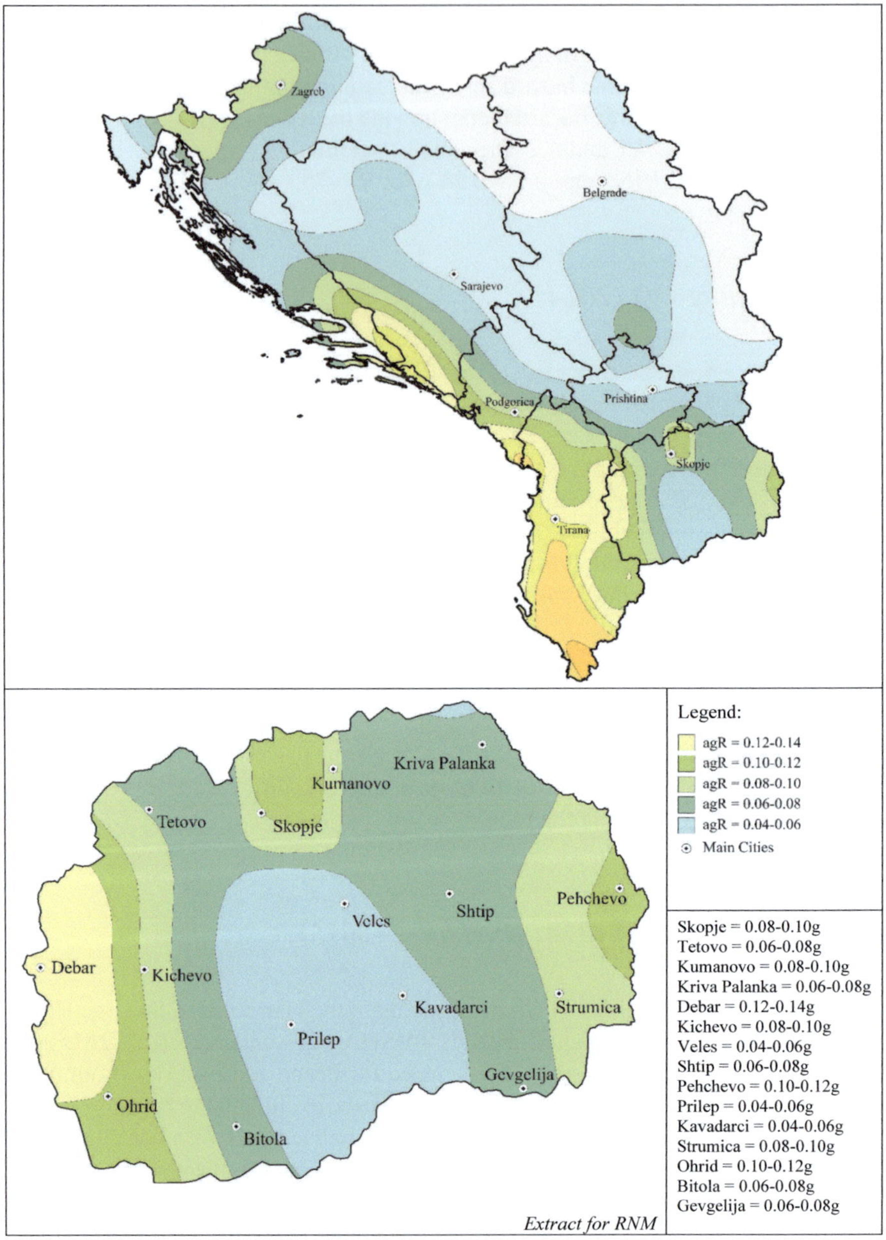

Fig. 4.1 BSHAP-1 seismic hazard map for return period of 95 years

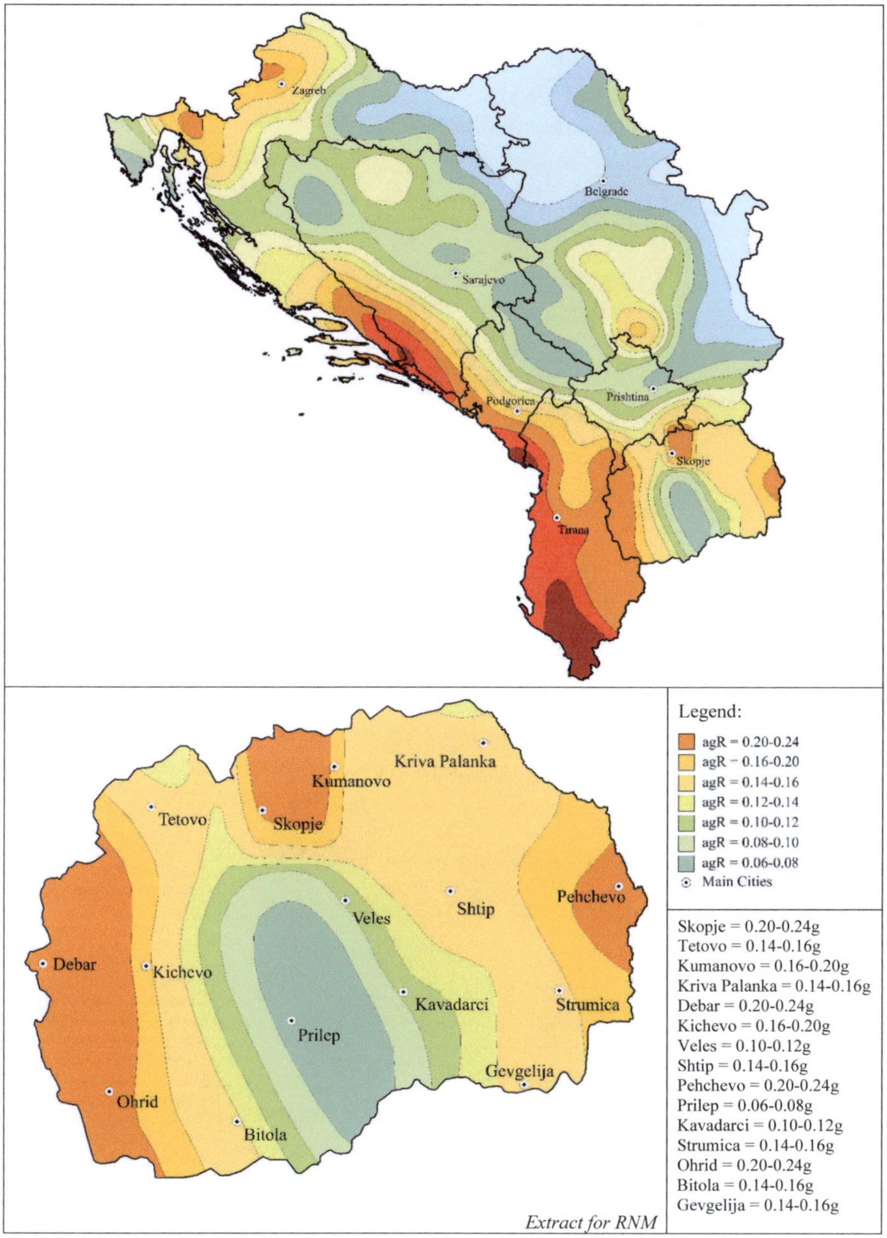

Fig. 4.2 BSHAP-1 seismic hazard map for return period of 475 years

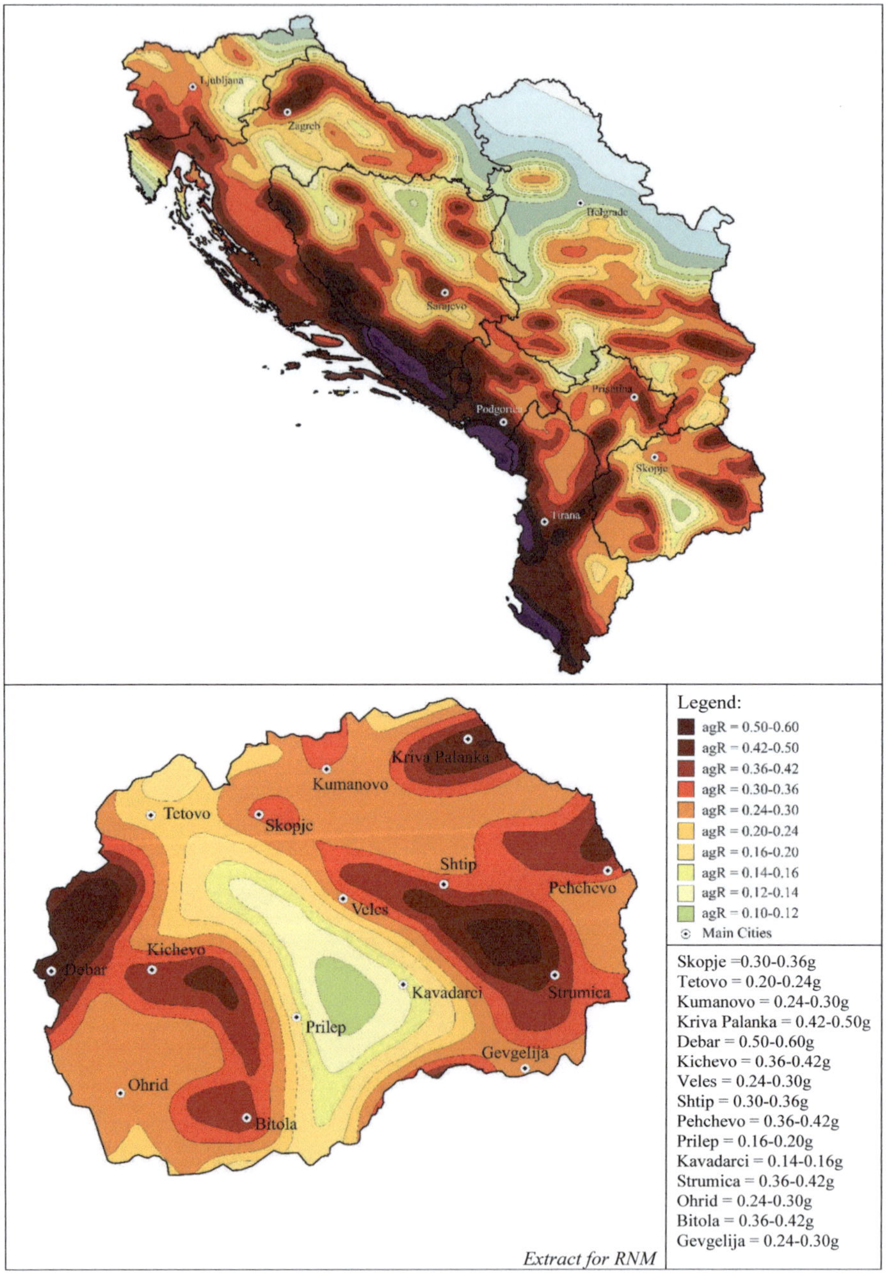

Fig. 4.3 BALKANS-OQ seismic hazard map for return period of 475 years

The probabilistic seismic hazard maps showing the RP95 and RP475 a_{gR} contours are presented Figs. 4.4 and 4.5. RP475 contoured a_{gR} values for RNM range from 0.04–0.16 g for RP95 and 0.08–0.30 g.

4.2 Global (European/Worldwide) Level Research Studies

4.2.1 GSHAP (1999)

Global Seismic Hazard Assessment Program (GSHAP) was UN/IDNDR Demonstration Project launched in 1992 by the International Lithosphere Program (ILP) and implemented in 1992–1999. The main objective of this project was to promote a regionally coordinated and homogeneous approach to seismic hazard evaluation at a worldwide level. The final map refers to PGA with 10% of probability in 50 years (475-year return period) and integrates the results obtained from the seven major regional areas (1—Central–North America, 2—South America, 3—Central–Northern Europe, 6—Middle East, 7—Northern Eurasia, 8—Eastern Asia, and 10—South-West Pacific) and 6 test areas (Northern Andes (PILOTO), Caucasus (CAUCAS), Adriatic Sea (ADRIA), East African Rift, India-China-Tibet-Myanmar-Bangla Dash, Iberia-Maghreb, and DACH). It should be mentioned that GSHAP was in relation to several cooperating projects: Mexico-C. America-Caribbean-S. America (PAIGH-IDRC), Circum Pannonian Basin (EU-QSEZ-CIRPAN), Eastern Mediterranean (RELEMR, USGS/UNESCO), Mediterranean (SESAME, IGCP-382), and Turkey and neighboring countries (Giardini 1999).

The results for RNM presented as an extract from the global map show values ranging from 0.12–0.42 g (Fig. 4.6). The presented map is developed in GIS form grid data freely available at http://gmo.gfz-potsdam.de/pub/download_data/download_data_frame.html.

4.2.2 SESAME (2000)

Seismotectonic and Seismic Hazard Assessment of the Mediterranean Basin (SESAME) is the IGCP-382 Project realized in the period 1996–2000, focused on the extension of the uniform seismic source zone model derived within GSHAP for Europe north of the Mediterranean (GSHAP Region 3) to the south. This project proposed a more detailed and integrated source model and homogeneous hazard for the Mediterranean region. This project follows a probabilistic approach based on the developed regional model of seismic sources with associated parameters based on selected most suitable attenuation relationships (Jimenez et al. 2003).

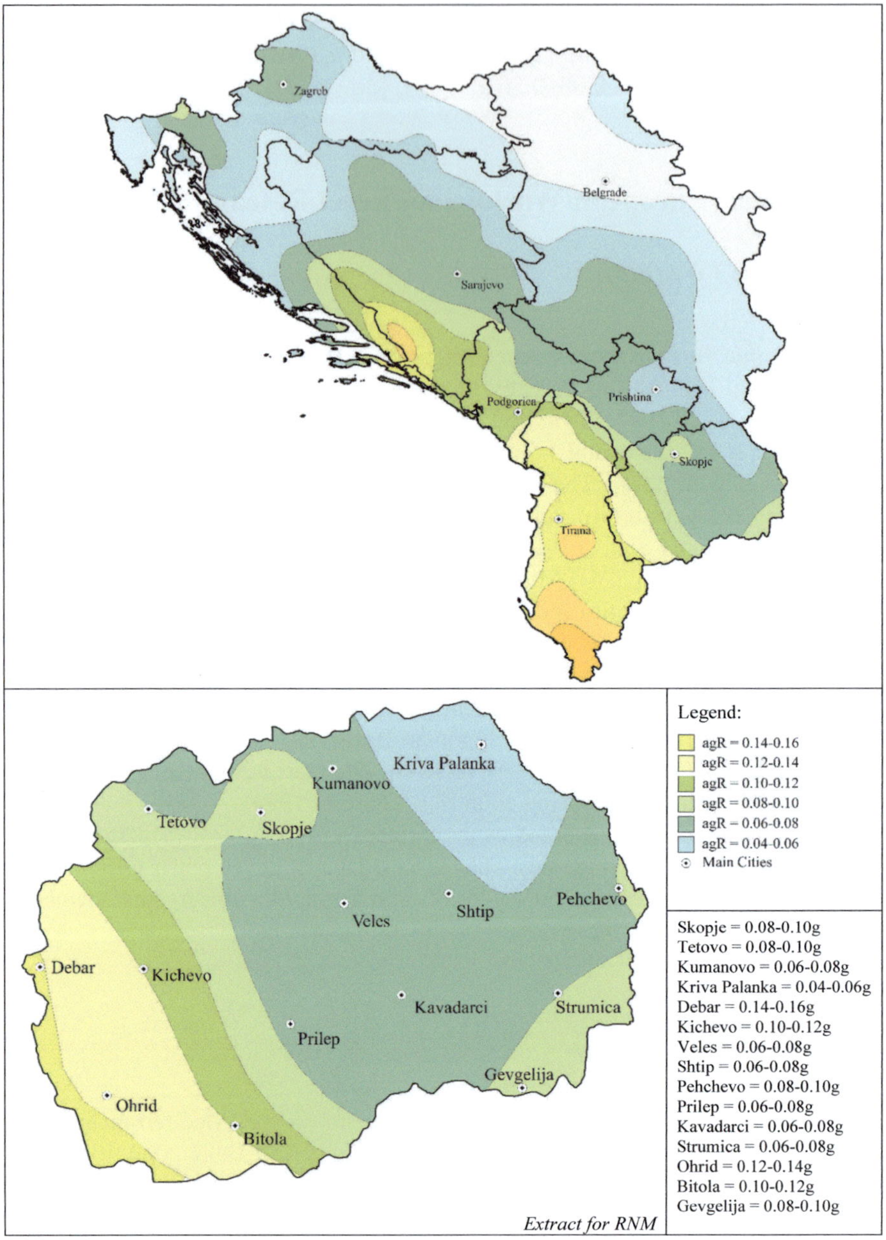

Fig. 4.4 BSHAP-2 seismic hazard map for return period of 95 years

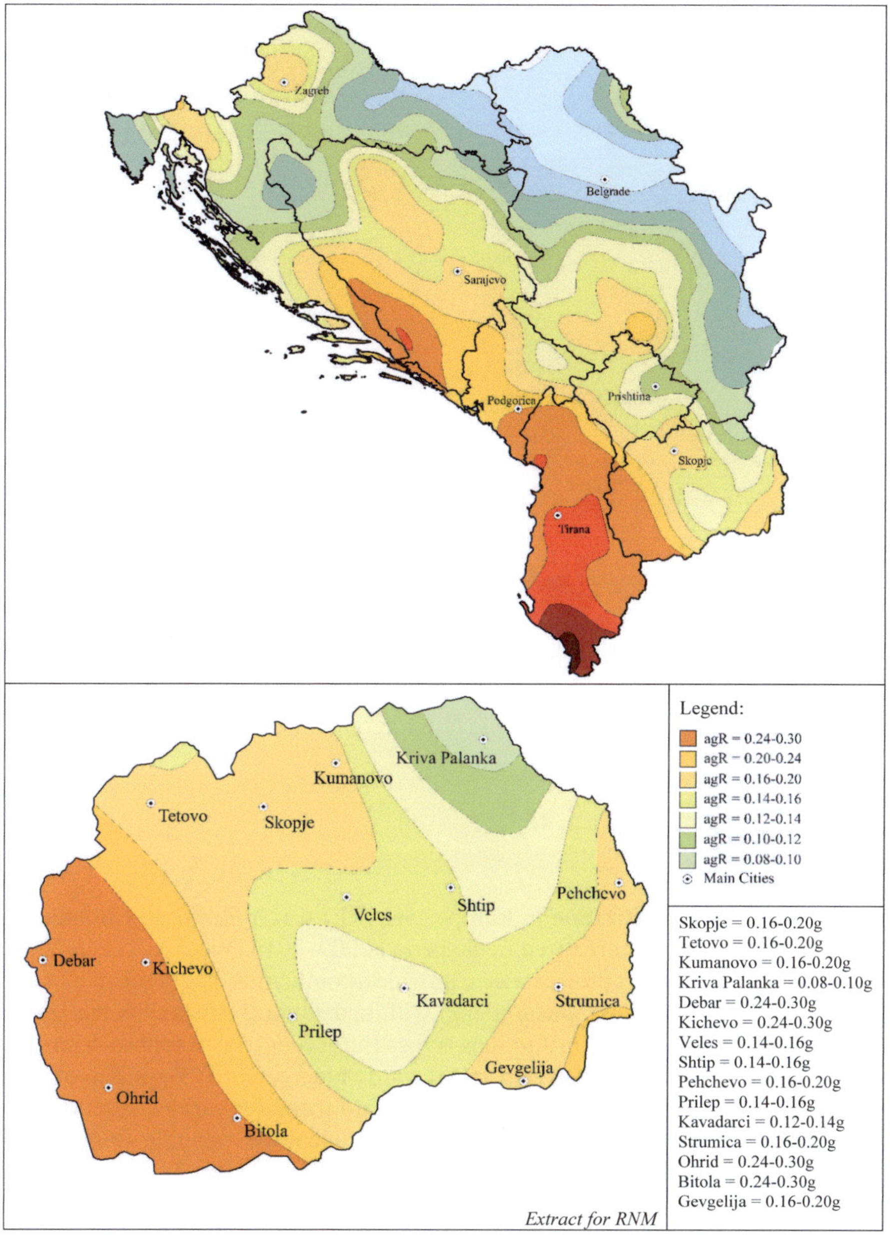

Fig. 4.5 BSHAP-2 seismic hazard map for return period of 475 years

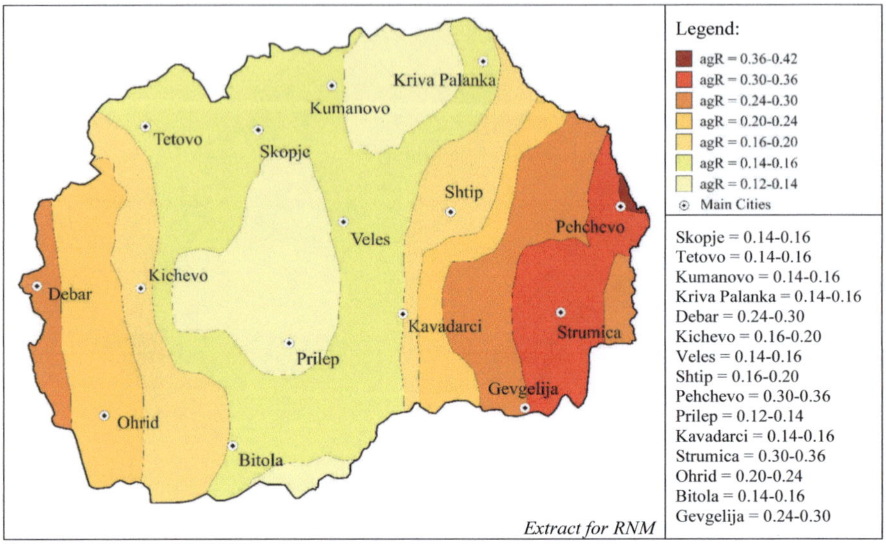

Fig. 4.6 GSHAP seismic hazard map for return period of 475 years (recolored and recontoured)

The results of SESAME could not be reliably deduced for RNM. Available maps are of low resolution and cannot be georeferenced or reproduced. Despite all the efforts from authors, associated seismic hazard grid data could not be found as well.

4.2.3 *ESHM (2013)*

Seismic Hazard Harmonization in Europe (SHARE) was a European Commission FP-7 collaborative project realized in the period 2009–2013 (Woessner et al. 2015). The main objective of this project was to provide a community-based seismic hazard model for the Euro-Mediterranean region with update mechanisms. Its realization was supported by 18 partner institutions across Europe and the Mediterranean.

The 2013 Euro-Mediterranean Seismic Hazard Model (ESHM13) consists of more than sixty-time-independent ground motion hazard maps for various spectral ordinates from PGA to 4 s and annual exceedance probabilities ranging from 10^{-1} to 10^{-4}. The hazard values are referenced to a rock velocity of $v_{s30} = 800$ m/s. ESHM13 model consists of earthquakes modeled as area sources, active faults, and smoothed background seismicity and includes all events with magnitudes $M_W \geq 4.5$ in the computation of hazard values. ESHM13 introduces an innovative weighting scheme that reflects the importance of the input datasets considering their time horizon, thus emphasizing the geologic knowledge for products with longer time horizons and seismological data for shorter ones (http://www.share-eu.org/node/6.html).

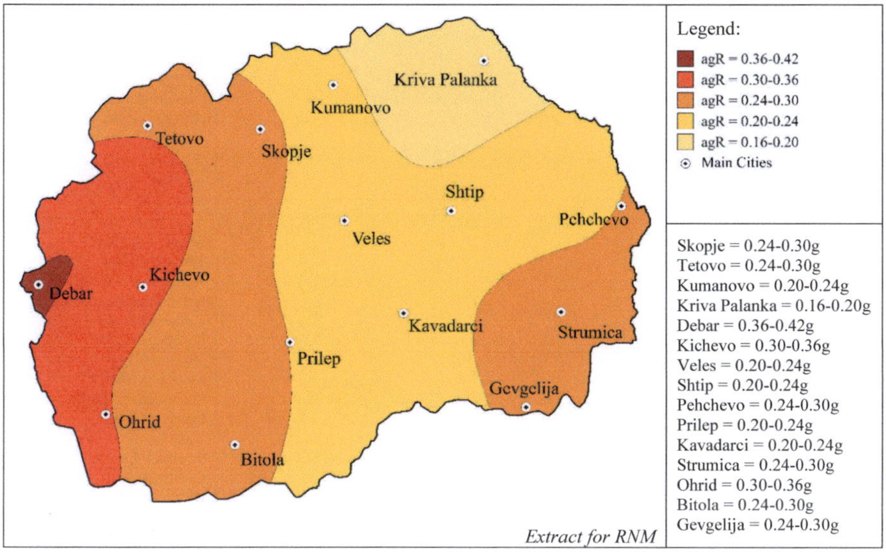

Fig. 4.7 ESHM13 seismic hazard map for return period of 475 years (recolored and recontoured)

The results for RNM presented as an extract from the SHARE map show values ranging from 0.16–0.42 g (Fig. 4.7). The presented map is developed in GIS form based on grid data freely available at http://www.efehr.org/en/Documentation/web-services/hazard-map-data/.

4.2.4 ESHM (2020)

The 2020 European Seismic Hazard Model (ESHM20) offers an updated assessment of earthquake hazard in the Euro-Mediterranean region (Danciu et al. 2021). Funded by the European Union's Horizon 2020 research and innovation program under grant agreement No. 730900 of the SERA-Project (www.sera-eu.org), ESHM20 follows the 2013 Euro-Mediterranean Seismic Hazard Model (ESHM13).

Collaborative efforts by research teams across Europe have led to the development of ESHM20 through a participatory peer-reviewed process, in partnership with Global Earthquake Model Foundation (GEM) and European Plate Observing System (EPOS).

The hazard model employs a fully probabilistic framework, ensuring that all datasets and inputs are harmonized across borders. The scientific data and input models are available in the publicly accessible EFEHR GitLab repository (https://gitlab.seismo.ethz.ch/efehr).

Of particular importance are also the two primary products that will form the basis for updating the informative annex of Eurocode 8, Part 1, as introduced by

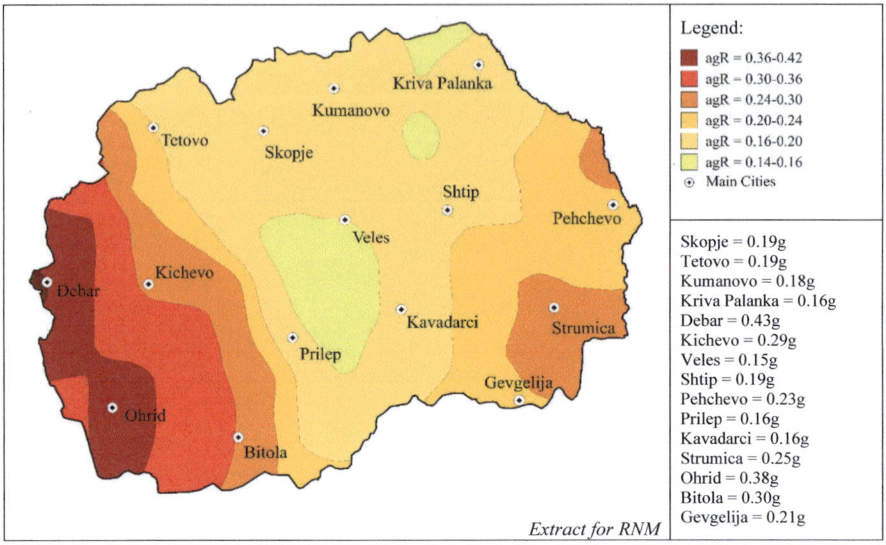

Fig. 4.8 ESHM20 seismic hazard map for return period of 476 years (recolored and recontoured)

CEN-TC250-SC8 regarding seismic action. The standard spectrum will be anchored by its plateau value, S_α, and its value at a 1-s spectral period, S_β, rather than the PGA. The informative annex will introduce two European hazard maps, one for S_α and one for S_β, both defined for a 475-year return period and based on the ESHM20 hazard model (Labbé and Paolucci 2022).

The results for RNM presented as an extract from the ESHM20 map show values ranging from 0.14–0.42 g (Fig. 4.8). The presented map is developed in GIS form based on grid data freely available at http://www.efehr.org/en/Documentation/web-services/hazard-map-data/.

References

Akkar S, Bommer JJ (2010) Empirical equations for the prediction of PGA, PGV and spectral accelerations in Europe, the Mediterranean region and the Middle East. Seismol Res Lett 81:195–206

Akkar S, Glavatovic B, Hoxha I, Kuk V, Zoranic A, Garevski M, Kovacevic S (2011) Harmonization of seismic hazard maps for the Western Balkan countries (BSHAP-1), NATO SfP-983054

Akkar S, Sandıkkaya MA, Bommer JJ (2014) Empirical ground-motion models for point- and extended-source crustal earthquake scenarios in Europe and the Middle East. Bull Earthquake Eng 12:359–387. https://doi.org/10.1007/s10518-013-9461-4

Bindi D, Luzi L, Massa M et al (2009) Horizontal and vertical ground motion prediction equations derived from the Italian Accelerometric Archive (ITACA). Bull Earthq Eng 8:1209–1230. https://doi.org/10.1007/s10518-009-9130-9

Bindi D, Massa M, Ameri G, Pacor F, Puglia R, Augliera P et al (2014) Pan-European ground-motion prediction equations for the average horizontal component of PGA, PGV, and 5 %-damped PSA at spectral periods up to 3.0 s using the RESORCE dataset. Bull Earthq Eng 12. https://doi.org/10.1007/s10518-013-9525-5

Boore DM, Atkinson GM (2008) Ground-motion prediction equations for the average horizontal component of PGA, PGV, and 5%-damped PSA at spectral periods between 0.01s and 10.0s. Earthq Spectra 241:99–138. https://doi.org/10.1193/1.2830434

Boore D, Stewart J, Seyhan E, Atkinson G (2014) NGA-West2 equations for predicting PGA, PGV, and 5% damped PSA for shallow crustal earthquakes. 30:1057–1085. https://doi.org/10.1193/070113EQS184M

Cauzzi C, Faccioli E (2008) Broadband (0.05 to 20 s) prediction of displacement response spectra based on worldwide digital records. J Seismol 12 (in press). https://doi.org/10.1007/s10950-008-9098-y

Chiou B, Youngs R (2014) Update of the Chiou and Youngs NGA model for the average horizontal component of peak ground motion and response spectra. Earthq Spectra 30:1117–1153. https://doi.org/10.1193/072813EQS219M

Danciu L, Nandan S, Reyes C, Basili R, Weatherill G, Beauval C, Rovida A, Vilanova S, Sesetyan K, Bard P-Y, Cotton F, Wiemer S, Giardini D (2021) The 2020 update of the European seismic hazard model: model overview, EFEHR Technical Report 001, v1.0.0. https://doi.org/10.12686/a15

Giardini D (1999) The global seismic hazard assessment program (GSHAP)-1992/1999. Annali di Geofisica 42(6). https://doi.org/10.4401/ag-3780

Gulerce Z, Salic R, Kuka N, Markusic S, Mihaljevic J, Kovacevic S (2015) Improvements in the harmonized seismic hazard maps for the Western Balkan countries (BSHAP-2), NATO SfP-984374

Jimenez MJ, Giardini D, Grünthal G (2003) The ESC-SESAME unified hazard model for the European-Mediterranean region. EMSC/CSEM Newsl 19:2–4

Labbé P, Paolucci R (2022) Developments relating to seismic action in the Eurocode 8 of next generation. In: Vacareanu R, Ionescu C (eds) Progresses in European earthquake engineering and seismology. ECEES 2022. Springer proceedings in earth and environmental sciences. Springer, Cham. https://doi.org/10.1007/978-3-031-15104-0_2

Pagani M, Monelli D, Weatherill G, Danciu L, Crowley H, Silva V, Henshaw P, Butler Nastasi L, Panzeri L, Simionato M, Vigano D (2014) OpenQuake engine: an open hazard (and risk) software for the global earthquake model. Seismol Res Lett 85(3):692–702. https://doi.org/10.1785/0220130087

Sevilgen V, Bennett RA, Brlek I, Danciu L, Kastelic V, Kovacevic S, Kreemer C, Kuk K, Kuka N, Milutinovic Z, Mustafa S, Sket-Motnikar B, Stein RS, Vucic L (2014) BALKANS-OQ: a collaborative seismic hazard assessment of the Balkan countries using the OpenQuake software and the GEM Strain Rate Model. In: Second European conference on earthquake engineering and seismology, Istanbul, 25–29 August 2014

Woessner J, Laurentiu D, Giardini D, Crowley H, Cotton F, Grünthal G, Valensise G, Arvidsson R, Basili R, Demircioglu MB, Hiemer S, Meletti C, Musson RW, Rovida AN, Sesetyan K, Stucchi M, The SHARE Consortium (2015) The 2013 European seismic hazard model: key components and results. Bull Earthq Eng 13:3553–3596. https://doi.org/10.1007/s10518-015-9795-1

Zabukovec B, Kuka N, Sostaric M, Motnikar BS, Suler T (2007) OHAZ: computer program for seismic hazard calculation, User Manual, Environmental Agency of Slovenia and Institute of Seismology of Albania, 65p

Chapter 5
Discussion and Conclusions

Abstract This chapter synthesizes the implications of the historical evolution of seismic design regulations and zoning maps for exposure, vulnerability, and seismic risk modeling in North Macedonia. It emphasizes that differences in code concepts and associated seismic zonation's—sometimes even between maps accompanying the same code—have caused significant spatial and temporal migration of design seismicity across the national territory. Analysis of intensity-zone coverage shows that, over successive official maps, the area assigned to low seismicity ($I_{MCS} \leq$ VI) has decreased from about one-third of the territory to less than 6%, while high-intensity zones (VIII–IX) have expanded or shifted, leaving only a small fraction of the country effectively outside seismic design requirements. The chapter introduces design and construction eras (pre-code, low-, moderate-, and high-code, plus Eurocode) based on the joint enforcement of seismic, concrete, and reinforced-concrete standards, which are crucial for constructing realistic exposure models. Comparative analyses of the base-shear coefficient *K* for typical masonry and RC buildings across maps and codes illustrate how changes in zoning and design philosophy translate into different seismic demands and safety levels. The chapter concludes that a national-scale seismic risk map is still lacking, mainly due to the absence of a comprehensive, time-dependent exposure model, and highlights the relevance of the presented synthesis for all SFRY successor states.

Keywords Seismic risk · Design code eras · Seismicity migration · Base-shear coefficient · Exposure modeling · Seismic vulnerability · SFRY successor countries

Of particular interest for exposure, seismic vulnerability, and seismic risk modelers are differences resulting from different concepts of building codes and differences arising from official seismic zonings accompanying different codes or even the same code (JUS31/81). Section 2 presents all official seismic zoning maps linked to building codes enforced from 1948 to 2020 (Table 1.1).

R. Salic Makreska and Z. Milutinovic, *Seismic Hazard Zonation for the Republic of North Macedonia*, SpringerBriefs in Geography,
https://doi.org/10.1007/978-3-032-13945-0_5

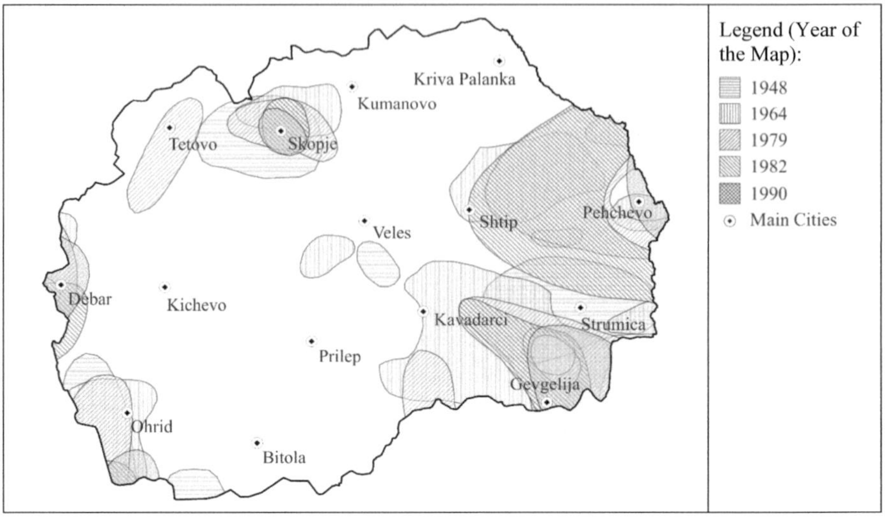

Fig. 5.1 Example of the spatial distribution of intensity zone IX in official seismic zonation

Evidently, seismic zones migrated spatially by every subsequent seismic zoning. Figure 5.1 depicts spatial migration of I_{MCS} = IX zone that from 1948 to 1982 incorporated 13.9% to a maximum 23.0% of the national territory, Table 5.1. According to RP475 SZM1990 the I_{MCS} = IX area drops to 5.86%.

The timeline of percent changes of the national territory by MSC zones is presented in Fig. 5.2. I_{MCS} = 10+ zone existed only in the period 1979–1982, viz., in Republic SZM1979. The $I_{MCS} = IX$ coverage, reaching its maximum in SZM1964 (22.96%), Table 5.1, gradually decreased to 5.67% with SZM1990. The sudden jump to 42.22% of the otherwise stable area of I_{MCS} = VII zone (22–24%) caused the sudden drop of I_{MCS} = VIII zone to 32.89%. At present, until the year 2024, the largest part of the country territory belongs to I_{MCS} = VIII zone (69.86%), then I_{MCS} = VII (22.41%), and I_{MCS} = IX (5.67%). The area of low seismicity, $I_{MCS} \leq$ VI, permanently decreased from 31.42% (SZM1948) to 2.06% (SZM1990). Thus, according to JUS31/81, only 2.06% of the country's territory is excluded from the earthquake-resistant design.

EN 1998-1 defines zones of very low seismicity ($a_{gR} \leq 0.04$ g) and low seismicity ($a_{gR} \leq 0.08$ g). For very low seismicity zone EN 1998-1 prescribes that the provisions of EN 1998 need not be observed. For low seismicity cases, allowed are reduced or simplified seismic design procedures for certain types or categories of structures. The lowest a_{gR} zone in RNM according to NCR SZM2020 is $0.10g < a_{gR} \leq 0.15g$, Fig. 2.8, encompassing 7.23% of the country territory, Table 5.1. Accordingly, the MKS EN 1998-1 provisions should be observed over the entire country's territory.

For 14 selected major towns and the City of Skopje, Table 5.2 quantifies the migration of I_{MCS} seismicity by seismic zoning maps. The most consistent zoning in the period 1948 to 1990 is for Skopje (IX–X, IX). One I_{MCS} degree migration is

Table 5.1 Percentage of the total country area in respect to official seismic zoning parameter and corresponding code

Code	SZM	Intensity degree							
		X	IX	VIII	VII	VI	V	IV	III
JUS61/48	1948	13.94		30.58	24.06	31.42			
JUS39/64	1964	-	22.96	11.85	22.72	22.55	0.18	15.67	4.07
	1979	1.94	19.93	53.17	24.96	-	-	-	-
JUS31/81	1982	2.55	15.88	32.89	42.22	6.46	-	-	-
	1990	-	5.67	69.86	22.41	2.06	-	-	-

MKS EN 1998-1/NA		$a_{gR}(g)$					
		0.30	**0.25**	**0.20**	**0.15**	**0.10**	**0.05**
NCR2020	≥ 2020	7.35	14.14	30.71	40.57	7.23	
DLR2020					5.91	50.73	43.36

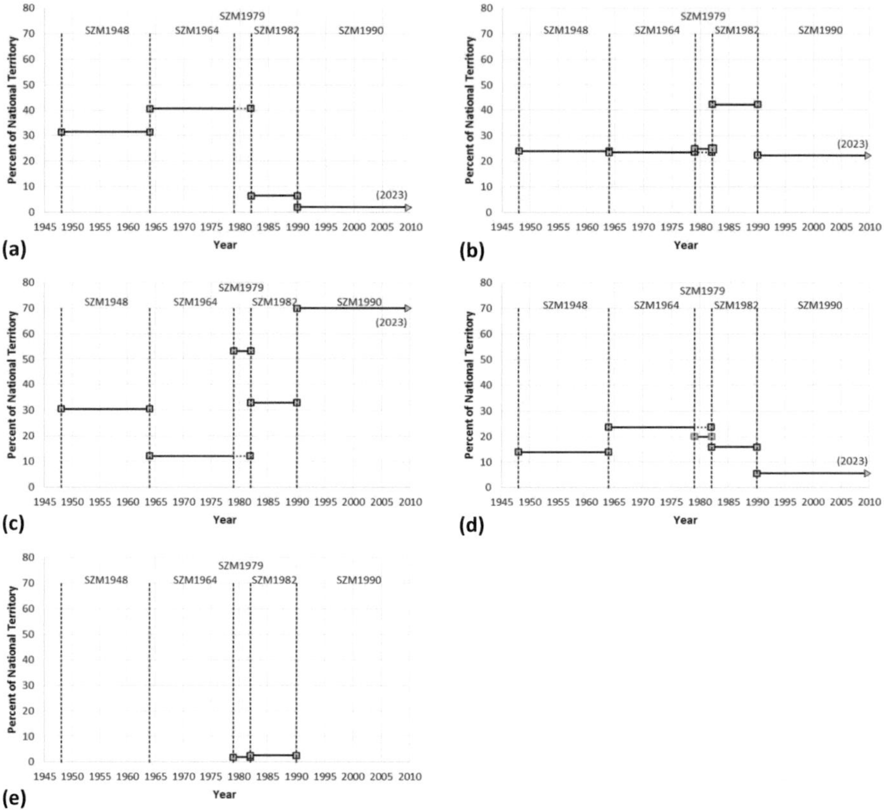

Fig. 5.2 Change over time in national territory coverage (%) by seismic intensity level (a) I_{MCS} = VI; (b) I_{MCS} = VII; (c) I_{MCS} = VIII; (d) I_{MCS} = IX; (e) I_{MCS} = X

identified for towns Kumanovo, Kriva Palanka, Debar, Veles, Shtip, Strumica Ohrid, Bitola, and Gevgelija, whereas for towns of Tetovo (VII to IX), Kichevo (VI–VII), Pehchevo (VIII–X), Kavadarci (VII–IX), and Prilep (VI–VIII) there is two I_{MCS} degrees migration. Kichevo and Prilep are very characteristic cases. The initially (in 1948) assigned seismicity dropped to I_{MCS} = VI; for Kichevo in the period 1964–1982; and Prilep in the period 1982–1990. According to JUSs in effect, no seismic design has been requested for both towns in referred periods. These two examples raise the question of the level of seismic protection and safety of earthquake not resistant construction. It is believed that some other larger human settlements in RNM are in the same position.

For 15 towns including City of Skopje, Table 5.3 presents the statistics (minimum, maximum, mean, median, and coefficient of variation, CV, values) of discussed seismic intensity (Fig. 5.3) and PGA (Fig. 5.4) seismic hazard/zoning maps. CV for the City of Skopje is the most minor (0.04). For towns with one-degree I_{MCS} migration CV ranges from 0.05 to 0.08. It is in range CV = 0.09–0.11 for towns

Table 5.2 Official zonation values for major cities in RNM

Zoning map	1948	1964	1979	1982	1990	2020
Town/city	I_{MCS} intensity degree					(a_{gR} in g)
1. Skopje	IX–X	IX	IX	IX	IX	0.25
2. Tetovo	VIII	VII	IX	VIII	VIII	0.20
3. Kumanovo	VII	VIII	VIII	VIII	VIII	0.20
4. Kriva Palanka	VII	VII	VIII	VIII	VIII	0.15
5. Debar	VIII	VIII	IX	IX	IX	0.30
6. Kichevo	VII	VI	VIII	VII	VIII	0.20
7. Veles	VIII	VII	VIII	VII	VIII	0.15
8. Shtip	VIII	IX	VIII	VIII	VIII	0.15
9. Pehchevo	VIII	IX	X	X	IX	0.30
10. Prilep	VII	VIII	VII	VI	VII	0.10
11. Kavadarci	VIII	IX	VIII	VII	VIII	0.15
12. Strumica	IX–X	VIII	VIII	VIII	VIII	0.25
13. Ohrid	IX–X	IX	IX	VIII	VIII	0.30
14. Bitola	VIII	VIII	VIII	VII	VIII	0.15
15. Gevgelija	VIII	IX	VIII	IX	VIII	0.20

with two-degree I_{MCS} migration. For Prilep and Kichevo CVs are 0.09 and 0.11. PGA-based CVs range from 0.26 to 0.55 for rock and 0.12 to 0.71 for ground type "characteristic soil," Table 5.3.

Also, of particular interest for authors of RP475 MKC EN 1998-1/NA:2020 (NCR SZM2020) seismic zoning map of RNM was to compare it with the SZMs of neighboring countries, Fig. 5.5.

At the northern part of the border with Bulgaria (BG), BG zone 0.15 g faces MK zone 0.15 g. At the central part of the border, BG zone 0.23 g faces the MK zones 0.15 g and 0.20 g partially. At the southern part of the border, BG zone 0.23 g faces the MK zones of 0.20 g and 0.25 g. At the border with Greece (GR), the GR zone 0.16 g faces MK zones of 0.20 g (west), 0.15 g (east and west from zone 0.10 g), and 0.10 g in the central part. The GR 0.24 g zone faces MK zones of 0.20 g and 0.25 g. With Serbia (RS), the MK/RS zones of 0.15 g and 0.20 g are in almost perfect agreement.

At the northern part of the border with Albania (AL),[1] AL zone 0.20 g faces MK zones 0.20 g and 0.25 g. In the central part, AL zone 0.25 g faces MK zones of 0.225 g and 0.30 g. At the southwestern part of the border, AL zone 0.15 g faces MK zone 0.30 g. North of zone 0.15 g, AL zone 020 g, on the north, faces MK zone of

[1] Kosovo and Albania have not yet officially adopted the Eurocodes. SZM map for Albania is taken from the Albanian draft Annex to EN 1998-1. The SZM of Kosovo was issued in 2009 by the Government of the Republic of Kosovo, Ministry of Energy and Mining. Since then, the design and construction practice has used it.

Table 5.3 Hazard assessment results, summary statistics

Town/city	Intensity I_{MCS}					PGA_{ROCK}					PGA_{GROUND}				
	I_{MCS}				[–]	g				[–]	g				[–]
	Min	Max	Mean	Median	CV	Min	Max	Mean	Median	CV	Min	Max	Mean	Median	CV
Skopje	9	10	9.17	9.00	0.04	0.15	0.35	0.25	0.22	0.29	0.15	0.45	0.30	0.30	0.71
Tetovo	7	9	8.17	8.00	0.09	0.14	0.38	0.23	0.22	0.27	0.28	0.45	0.36	0.36	0.34
Kumanovo	7	8	7.83	8.00	0.05	0.11	0.30	0.21	0.21	0.26	0.15	0.35	0.25	0.25	0.56
K. Palanka	7	8	7.67	8.00	0.07	0.07	0.46	0.19	0.16	0.52	0.15	0.25	0.20	0.20	0.36
Debar	8	9	8.67	9.00	0.06	0.22	0.58	0.35	0.32	0.31	0.28	0.55	0.41	0.41	0.47
Kichevo	6	8	7.33	7.50	0.11	0.17	0.50	0.27	0.25	0.32	0.28	0.35	0.31	0.31	0.17
Veles	7	8	7.67	8.00	0.07	0.10	0.27	0.17	0.15	0.31	0.15	0.35	0.25	0.25	0.56
Shtip	8	9	8.17	8.00	0.05	0.10	0.35	0.20	0.19	0.38	0.15	0.35	0.25	0.25	0.56
Pehchevo	8	10	9.17	9.00	0.08	0.17	0.85	0.36	0.29	0.55	0.28	0.55	0.41	0.41	0.47
Prilep	6	8	7.00	7.00	0.09	0.07	0.30	0.15	0.15	0.37	0.15	0.25	0.20	0.20	0.36
Kavadarci	7	9	8.00	8.00	0.08	0.09	0.35	0.18	0.16	0.40	0.15	0.35	0.25	0.25	0.56
Strumica	8	10	8.33	8.00	0.10	0.15	0.65	0.32	0.27	0.44	0.15	0.45	0.30	0.30	0.71
Ohrid	8	10	8.67	8.50	0.09	0.20	0.66	0.34	0.29	0.36	0.15	0.45	0.30	0.30	0.71
Bitola	7	8	7.83	8.00	0.05	0.13	0.46	0.24	0.24	0.37	0.15	0.25	0.20	0.20	0.36
Gevgelija	8	9	8.50	8.50	0.06	0.13	0.46	0.27	0.25	0.42	0.15	0.45	0.30	0.30	0.71

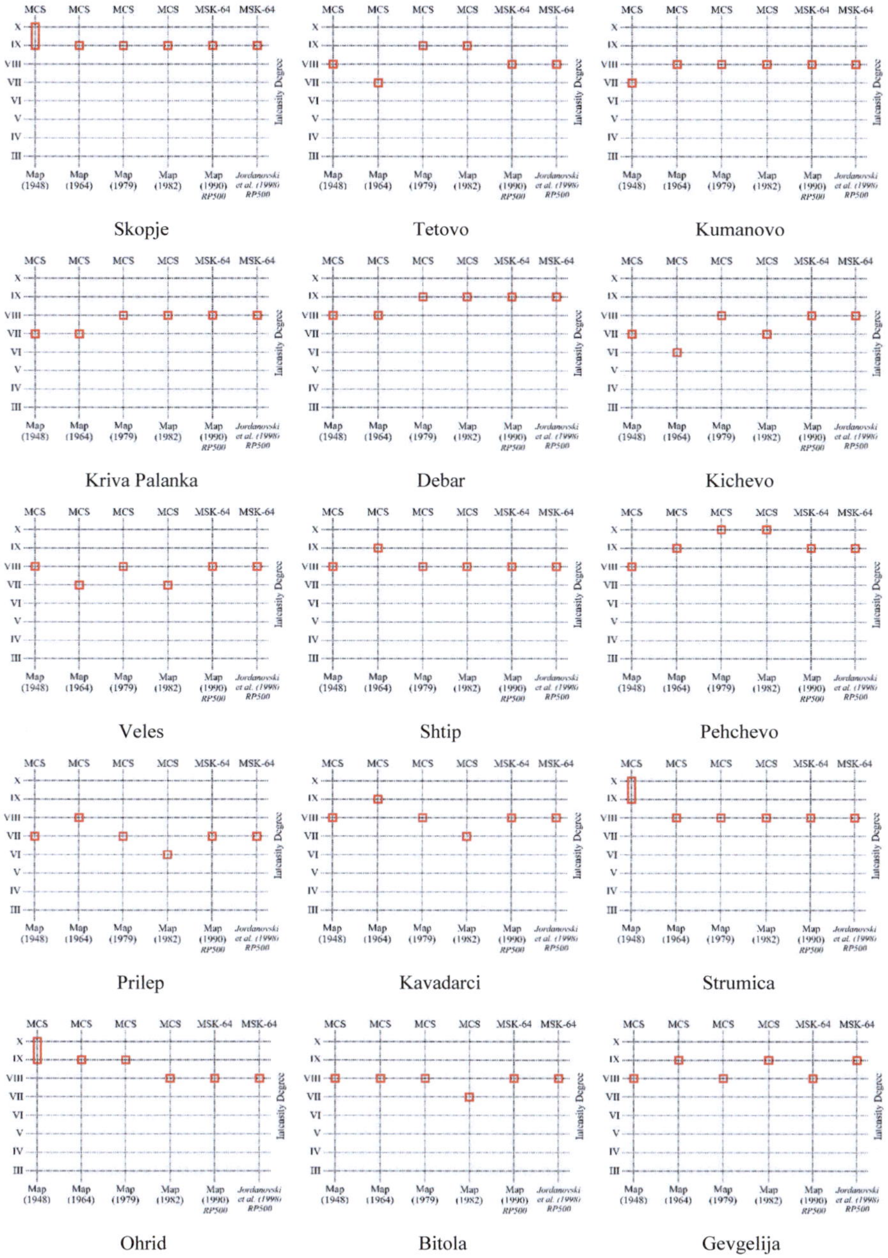

Fig. 5.3 Intensity values per selected cities due to different intensity maps

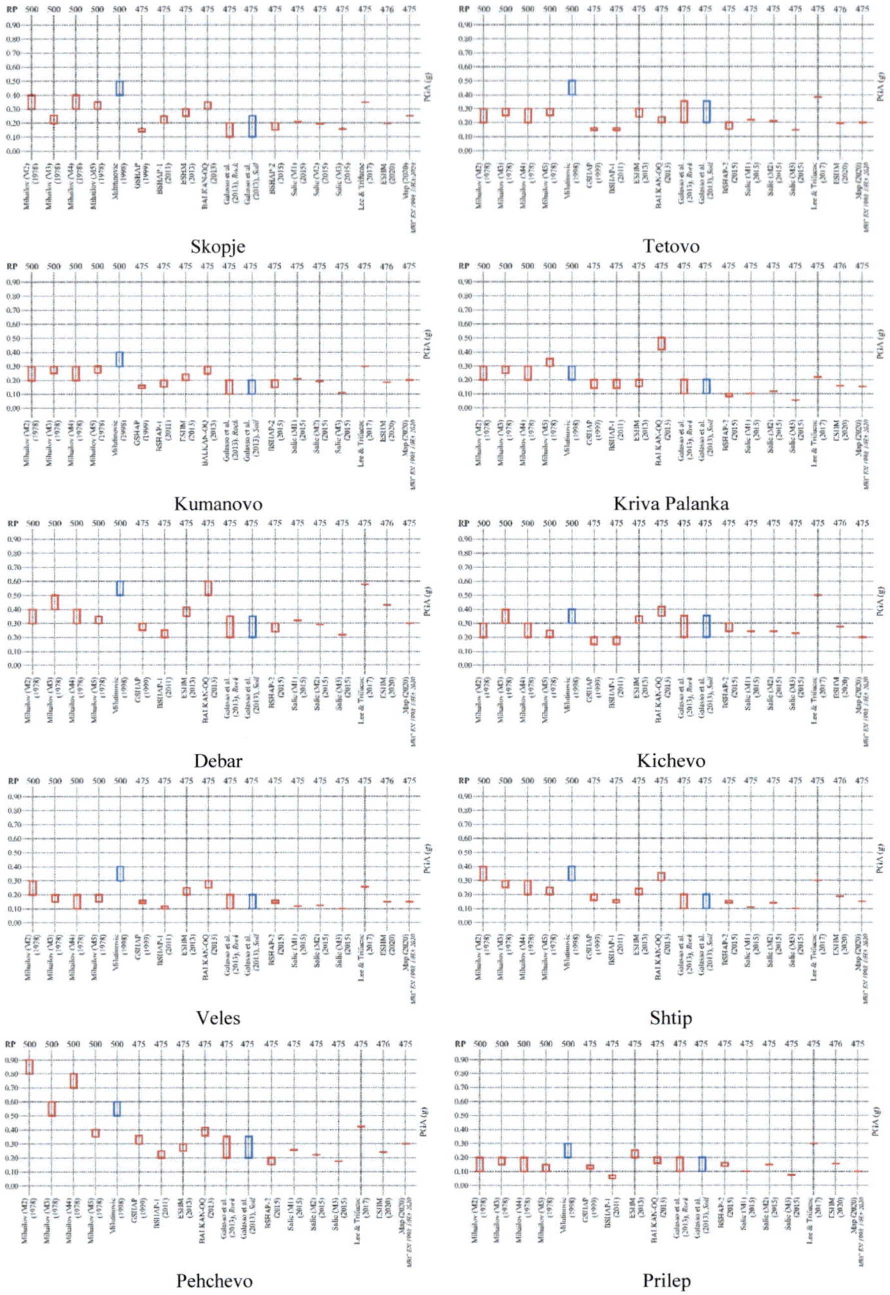

Fig. 5.4 Variation in PGA values for selected cities in North Macedonia according to different PGA maps

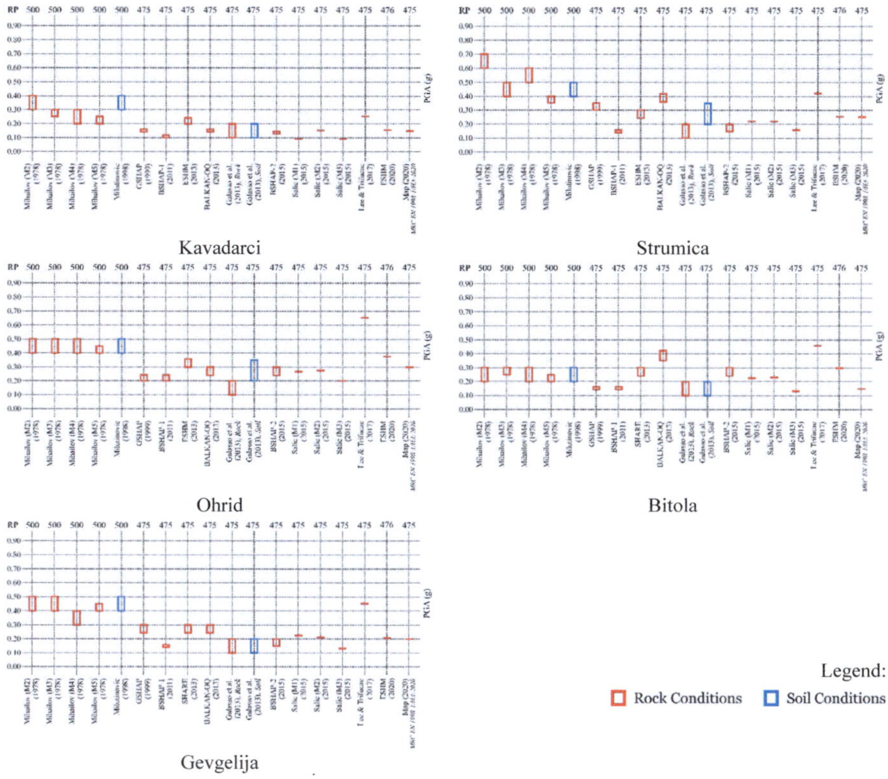

Fig. 5.4 (continued)

0.30 g, and 0.20 g, on the south. At the southwest part of the border with Kosovo (XK), the XK zone 0.22 g faces the MK zone 0.20 g. At the north part of the border, the XK zones 0.16 g and the 0.18 g confront the MK zone 0.20 g.

The best, almost perfect agreement is between Serbian and Macedonian seismic zonings presented in respective National Annexes (NA). The worst agreement of zones is with Albania reaching 0.15 g in Debar (MK) region.

Figure 5.6 summarizes all seismic zoning/hazard maps adopted in National Annexes of countries succeeding SFRY, Albania, Greece, and Bulgaria. The problem of the aggregated map is the equidistant step of contouring. Croatia, Bosnia and Herzegovina, Kosovo, and Montenegro contour their SZM maps by the step of 0.02 g. Its sims that these NA maps are direct seismic hazard analysis products presented as seismic zoning. Slovenia maps at 0.025 g. Albania, RNM, and Serbia map a_{gR} contours by 0.05 g. RNM seismic zoning (Milutinovic et al. 2016) includes a policy to map certain territories in a higher a_{gR} zones due to inaccessibility for response operations. Albanian and Serbian seismic zoning policies are not known. Greece contours at 0.08 g, whereas Bulgaria uses non-equidistant contouring (0.04, 0.08,

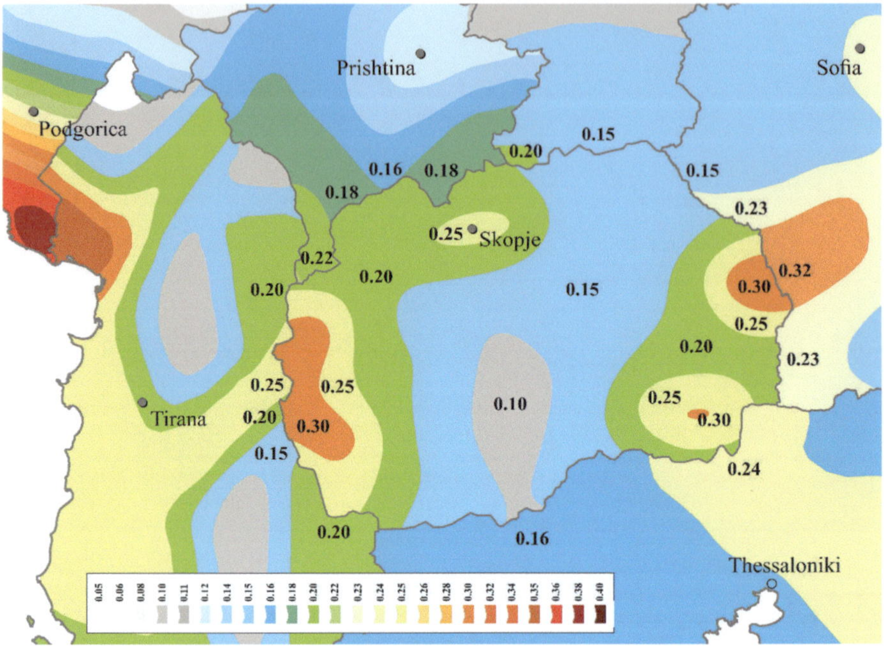

Fig. 5.5 RP475 RNM seismic zoning map versus RP475 maps presented in National Annexes of neighbor countries, by alphabetic order, Albania, Bulgaria, Greece, Kosovo, and Serbia

and 0.09 g). There are methods to, say, 0.16 or 0.26 g zones split along their central lines, but authors preferred to avoid reinterpretation of enacted seismic zonings.

To test and demonstrate the influence of seismic zoning maps and related building codes for N. Macedonia, Table 5.4 we studied the variation of the base shear coefficient (K) for regular typical buildings. This coefficient is defined as the dimensionless ratio of base shear to building weight. For comparative purposes, we selected a pilot building with an approximate base area of 600 m^2, utilizing different structural systems (masonry and reinforced concrete—RC). Masonry structures have total loads of 15–16 kN/m^2, with floor weights of 9500 kN for the ground floor and 6000 kN for the upper floors. RC structures have loads of 11–13 kN/m^2, with floor weights of 8000 kN for the ground floor and 5000 kN for the upper floors. We computed the K values for different periods, using various building codes and seismic maps, to observe its fluctuations in 14 towns in North Macedonia for both masonry and RC buildings (Table 5.4).

The analysis included masonry buildings with natural periods of 0.1, 0.2, and 0.3 s and RC buildings with periods ranging from 0.6–0.9 s. The results indicate that K is higher for masonry structures, making them more susceptible to seismic forces. Additionally, newer codes, such as the Eurocodes 2020, require higher K values, reflecting stricter safety standards.

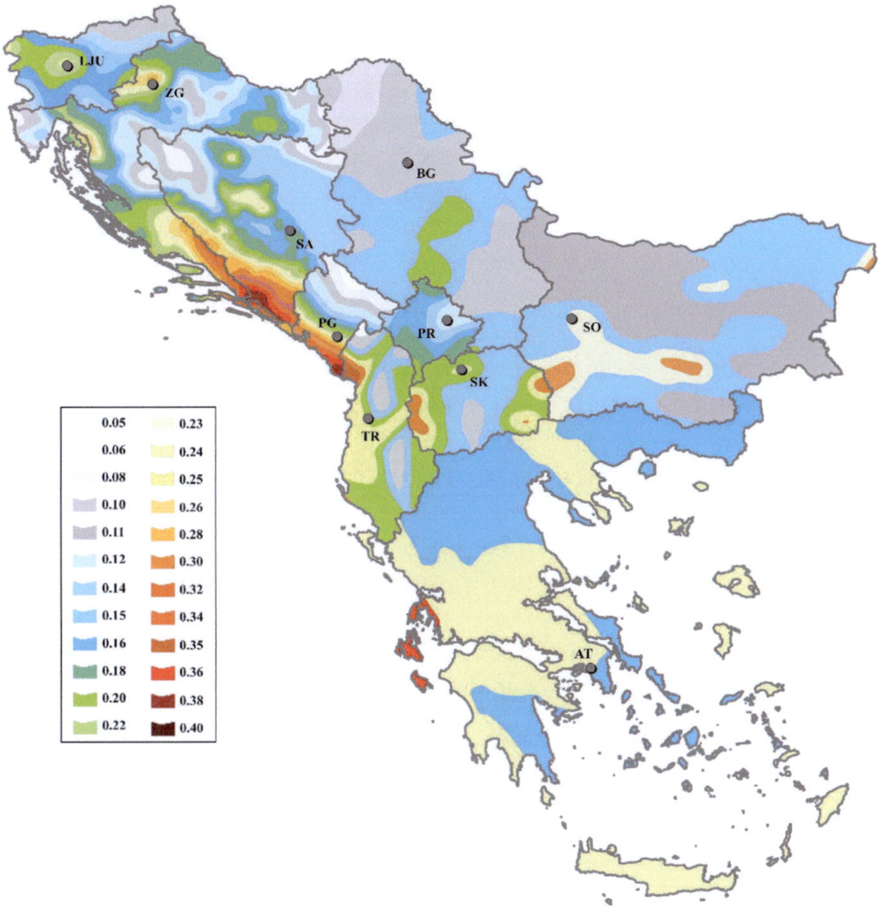

Fig. 5.6 RP475 RNM seismic zoning map vs. RP475 maps presented in National Annexes of, by alphabetic order, Albania, Bosnia and Herzegovina, Bulgaria, Croatia, Greece, Kosovo, Montenegro, Serbia, and Slovenia

In the absence of national studies quantifying differences between classic (earthquake non-resistant) and earthquake-resistant design, we cite the Harihara Venkata Nagasai et al. (2019) study results. The authors analyzed the GF + 7 (GF: ground floor) 24 m in height, 3D RC frame hostel building located in the seismic zone III (Z = 0.16) of India. Dead (DL), live (LL), and earthquake (EL) loads of the structure are defined as per IS 875-part1, 2, 3, and IS 1893, respectively. As per IS: 1893-2002, the authors checked the gravitational load combination $1.5 \cdot (DL \pm LL)$ against eight seismic design situation effects combinations, four corresponding to $E_d = 1.2 \cdot \left(DL + LL \pm EL_{x,y}\right)$ and another four corresponding to $E_d = 1.5 \cdot \left(DL \pm EL_{x,y}\right)$, to obtain the ratios between the "seismic"/"non-seismic support reactions" of 247.02/128.72 kNm (1.91 times) for bending moment, 269.12/

Table 5.4 Variation of K (%) in respect to design codes and corresponding SH maps

City	Design code	SH map	Masonry structure			RC structure			
			$T_1 =$ 0.1 s	$T_1 =$ 0.2 s	$T_1 =$ 0.3 s	$T_1 =$ 0.6 s	$T_1 =$ 0.7 s	$T_1 =$ 0.8 s	$T_1 =$ 0.9 s
Skopje	1948	1948	2.00	2.00	2.00	*	*	*	*
	1964	1964	11.88	10.80	10.24	8.05	6.90	5.97	5.25
		1979	11.88	10.80	10.24	8.05	6.90	5.97	5.25
	1981	1979	20.00	20.00	20.00	8.30	7.14	6.25	5.56
		1982	20.00	20.00	20.00	8.30	7.14	6.25	5.56
		1990	20.00	20.00	20.00	8.30	7.14	6.25	5.56
	2020	2020	41.70	41.70	35.40	9.10	7.82	6.80	7.10
Tetovo	1948	1948	1.50	1.50	1.50	*	*	*	*
	1964	1964	2.97	2.70	2.56	2.01	1.71	1.49	1.31
		1979	11.88	10.80	10.24	8.05	6.90	5.97	5.25
	1981	1979	20.00	20.00	20.00	8.30	7.14	6.25	5.56
		1982	10.00	10.00	10.00	4.15	3.57	3.13	2.78
		1990	10.00	10.00	10.00	4.15	3.57	3.13	2.78
	2020	2020	33.33	33.33	28.05	7.23	6.21	5.44	5.70
Kumanovo	1948	1948	1.00	1.00	1.00	*	*	*	*
	1964	1964	5.93	5.40	5.12	4.02	3.42	3.00	2.63
		1979	5.93	5.40	5.12	4.02	3.42	3.00	2.63
	1981	1979	10.00	10.00	10.00	4.15	3.57	3.13	2.78
		1982	10.00	10.00	10.00	4.15	3.57	3.13	2.78
		1990	10.00	10.00	10.00	4.15	3.57	3.13	2.78
	2020	2020	33.33	33.33	28.05	7.23	6.21	5.44	5.70
Kriva Palanka	1948	1948	1.00	1.00	1.00	*	*	*	*
	1964	1964	2.97	2.70	2.56	2.01	1.71	1.49	1.31
		1979	5.93	5.40	5.12	4.02	3.42	3.00	2.63
	1981	1979	10.00	10.00	10.00	4.15	3.57	3.13	2.78
		1982	10.00	10.00	10.00	4.15	3.57	3.13	2.78
		1990	10.00	10.00	10.00	4.15	3.57	3.13	2.78
	2020	2020	25.00	25.00	21.25	5.44	4.68	4.08	4.30
Debar	1948	1948	1.50	1.50	1.50	*	*	*	*
	1964	1964	5.93	5.40	5.12	4.02	3.42	3.00	2.63
		1979	11.88	10.80	10.24	8.05	6.90	5.97	5.25
	1981	1979	20.00	20.00	20.00	8.30	7.14	6.25	5.56
		1982	20.00	20.00	20.00	8.30	7.14	6.25	5.56
		1990	20.00	20.00	20.00	8.30	7.14	6.25	5.56
	2020	2020	50.00	50.00	42.50	10.88	9.35	8.16	8.50

(continued)

Table 5.4 (continued)

City	Design code	SH map	Masonry structure			RC structure			
			$T_1 =$ 0.1 s	$T_1 =$ 0.2 s	$T_1 =$ 0.3 s	$T_1 =$ 0.6 s	$T_1 =$ 0.7 s	$T_1 =$ 0.8 s	$T_1 =$ 0.9 s
Kichevo	1948	1948	1.00	1.00	1.00	*	*	*	*
	1964	1964	**	**	**	**	**	**	**
		1979	5.93	5.40	5.12	4.02	3.42	3.00	2.63
	1981	1979	10.00	10.00	10.00	4.15	3.57	3.13	2.78
		1982	5.00	5.00	5.00	2.08	1.79	1.56	1.39
		1990	10.00	10.00	10.00	4.15	3.57	3.13	2.78
	2020	2020	33.33	33.33	28.05	7.23	6.21	5.44	5.70
Veles	1948	1948	1.50	1.50	1.50	*	*	*	*
	1964	1964	2.97	2.70	2.56	2.01	1.71	1.49	1.31
		1979	5.93	5.40	5.12	4.02	3.42	3.00	2.63
	1981	1979	10.00	10.00	10.00	4.15	3.57	3.13	2.78
		1982	5.00	5.00	5.00	2.08	1.79	1.56	1.39
		1990	10.00	10.00	10.00	4.15	3.57	3.13	2.78
	2020	2020	25.00	25.00	21.25	5.44	4.68	4.08	4.30
Shtip	1948	1948	1.50	1.50	1.50	*	*	*	*
	1964	1964	11.88	10.80	10.24	8.05	6.90	5.97	5.25
		1979	5.93	5.40	5.12	4.02	3.42	3.00	2.63
	1981	1979	10.00	10.00	10.00	4.15	3.57	3.13	2.78
		1982	10.00	10.00	10.00	4.15	3.57	3.13	2.78
		1990	10.00	10.00	10.00	4.15	3.57	3.13	2.78
	2020	2020	25.00	25.00	21.25	5.44	4.68	4.08	4.30
Pehchevo	1948	1948	1.50	1.50	1.50	*	*	*	*
	1964	1964	11.88	10.80	10.24	8.05	6.90	5.97	5.25
		1979	17.82	16.20	15.37	12.07	10.26	8.95	7.88
	1981	1979	***	***	***	***	***	***	***
		1982	***	***	***	***	***	***	***
		1990	20.00	20.00	20.00	8.30	7.14	6.25	5.56
	2020	2020	50.00	50.00	42.50	10.88	9.35	8.16	8.50
Prilep	1948	1948	1.00	1.00	1.00	*	*	*	*
	1964	1964	5.93	5.40	5.12	4.02	3.42	3.00	2.63
		1979	2.97	2.70	2.56	2.01	1.71	1.49	1.31
	1981	1979	5.00	5.00	5.00	2.08	1.79	1.56	1.39
		1982	**	**	**	**	**	**	**
		1990	5.00	5.00	5.00	2.08	1.79	1.56	1.39

(continued)

Table 5.4 (continued)

City	Design code	SH map	Masonry structure			RC structure			
			$T_1 =$ 0.1 s	$T_1 =$ 0.2 s	$T_1 =$ 0.3 s	$T_1 =$ 0.6 s	$T_1 =$ 0.7 s	$T_1 =$ 0.8 s	$T_1 =$ 0.9 s
	2020	2020	16.70	16.70	14.95	3.66	3.15	2.72	2.80
Kavadarci	1948	1948	1.50	1.50	1.50	*	*	*	*
	1964	1964	11.88	10.80	10.24	8.05	6.90	5.97	5.25
		1979	5.93	5.40	5.12	4.02	3.42	3.00	2.63
	1981	1979	10.00	10.00	10.00	4.15	3.57	3.13	2.78
		1982	5.00	5.00	5.00	2.08	1.79	1.56	1.39
		1990	10.00	10.00	10.00	4.15	3.57	3.13	2.78
	2020	2020	25.00	25.00	21.25	5.44	4.68	4.08	4.30
Strumica	1948	1948	2.00	2.00	2.00	*	*	*	*
	1964	1964	5.93	5.40	5.12	4.02	3.42	3.00	2.63
		1979	5.93	5.40	5.12	4.02	3.42	3.00	2.63
	1981	1979	10.00	10.00	10.00	4.15	3.57	3.13	2.78
		1982	10.00	10.00	10.00	4.15	3.57	3.13	2.78
		1990	10.00	10.00	10.00	4.15	3.57	3.13	2.78
	2020	2020	41.70	41.70	35.40	9.10	7.82	6.80	7.10
Ohrid	1948	1948	2.00	2.00	2.00	*	*	*	*
	1964	1964	11.88	10.80	10.24	8.05	6.90	5.97	5.25
		1979	11.88	10.80	10.24	8.05	6.90	5.97	5.25
	1981	1979	20.00	20.00	20.00	8.30	7.14	6.25	5.56
		1982	10.00	10.00	10.00	4.15	3.57	3.13	2.78
		1990	10.00	10.00	10.00	4.15	3.57	3.13	2.78
	2020	2020	50.00	50.00	42.50	10.88	9.35	8.16	8.50
Bitola	1948	1948	1.50	1.50	1.50	*	*	*	*
	1964	1964	5.93	5.40	5.12	4.02	3.42	3.00	2.63
		1979	5.93	5.40	5.12	4.02	3.42	3.00	2.63
	1981	1979	10.00	10.00	10.00	4.15	3.57	3.13	2.78
		1982	5.00	5.00	5.00	2.08	1.79	1.56	1.39
		1990	10.00	10.00	10.00	4.15	3.57	3.13	2.78
	2020	2020	25.00	25.00	21.25	5.44	4.68	4.08	4.30
Gevgelija	1948	1948	1.50	1.50	1.50	*	*	*	*
	1964	1964	11.88	10.80	10.24	8.05	6.90	5.97	5.25
		1979	5.93	5.40	5.12	4.02	3.42	3.00	2.63
	1981	1979	10.00	10.00	10.00	4.15	3.57	3.13	2.78
		1982	20.00	20.00	20.00	8.30	7.14	6.25	5.56

(continued)

Table 5.4 (continued)

City	Design code		SH map	Masonry structure			RC structure			
				$T_1 =$ 0.1 s	$T_1 =$ 0.2 s	$T_1 =$ 0.3 s	$T_1 =$ 0.6 s	$T_1 =$ 0.7 s	$T_1 =$ 0.8 s	$T_1 =$ 0.9 s
			1990	10.00	10.00	10.00	4.15	3.57	3.13	2.78
		2020	2020	33.33	33.33	28.05	7.23	6.21	5.44	5.70

* RC buildings are not considered in the corresponding code
** For intensity VI no seismic design force is considered in the calculations
*** For intensity X no corresponding calculation parameters are given in the code

146.17 kN (1.84) for maximum shear force, and 7349.70/5909.08 kN (1.24) for axial force. Rewritten in percent, it is 47.9%, 45.7%, and 19.6% increase, respectively. The study compares top-story maximum <*x*, *y*, *z*> deflections estimated at 0.42/16.6 (39.5 times), 0.65/90.9 (139.8 times), and 22.5/76.2 mm (3.4), respectively. If, by chance, the analyzed building was located in seismic zone IV ($Z = 0.24$) or V ($Z = 0.36$), characteristic values of seismic actions, $EL_{x,y}$, should be multiplied by Z_{IV}/Z_{III} and Z_V/Z_{III} factors equalling 1.5 and 2.25, respectively. The increase of $EL_{x,y}$ will further cause the increase of componental values of design action effects, $E_{d,x}$.

All analyses were performed by SAP 2000 following Indian IS 1893 (Part-1):2002 Earthquake Design Code (R-format code) for Importance factor $\gamma_I = 1.0$, response reduction factor $R = R_p = q = 3$, and ground type medium (mostly black cotton) soil.

Design and construction in Kichevo and Prilep during the $I_{\mathrm{MCS}} =$ VI period were exclusively controlled by codes for concrete and reinforced concrete (CRC), nationally referred to as PBAB. The first CRC was enforced in 1947, the second in 1971, and the third in 1987. CRC46/47 (JUS46/47) and CRC51/71 (JUS51/71) did not distinguish reinforcement detailing for non-seismic and seismic regions. Both CRCs allowed maximum distances between stirrups in columns min $\{15\Phi, b_o\}$, where b_o is the smallest cross-section dimension and Φ is the diameter of the smallest longitudinal rebar. For non-seismic regions, CRC11/87 (JUS11/87) requests min $\{15\Phi, b_o, 30\,\mathrm{cm}\}$. To assure ductility of columns in seismic regions in top and bottom critical regions and rebar lap zone CRC11/87 request closed stirrups (full overlap over the whole length of the shorter side) spaced min $\{7.5\Phi, 10c\}$. Out of these zones, CRC11/87 request standard (135° bent) stirrups spaced min $\{15\Phi, 20cm\}$. In addition, JUS31/81 requests 50% rebar lapping outside the zones of plastic hinges (critical regions) and in areas where the tensile stresses are the least (about the central part of the column). Consequently, for standard story heights of 3.0 m, the stirrups are spaced, almost over the entire column height, according to min $\{7.5\Phi, 10cm\}$ condition, and very few according to min $\{15\Phi, 20cm\}$.

The towns of Kichevo and Prilep, according to SZM1990, are located in $I_{MCS} = VIII$ and $I_{MCS} = VII$ zones are typical examples of quite adverse influence of migration of seismic zones in consecutive seismic zonings that seriously affect the level of seismic protection of structures and the safety of inhabitants. A rigorous analysis of locations of human settlements in RNM should be conducted to identify and systemize those that, during their development, migrated to $I_{MCS} \leq VI$ design seismicity zones.

Codes control the resistance of exposure to adverse natural and social events. Rephrasing codes are the only efficient instrument that prescribes and controls the acceptable national risk level to a particular adverse natural or social event. Before adopting Eurocodes, the exposure in RNM and SFRY was developed according to three seismic design standards and three standards for concrete and reinforced concrete. Their enforcement was not temporally harmonized. Enforcement of CRC standards was delayed for six to seven years following the enforcement of seismic design standards.

For vulnerability assessment of the exposure, the year of design (optimally) or the year of construction (alternative available from AREC) is essential information for defining the era of codes used because they reflect significant changes in design forces or in detailing requirements that matter the seismic performance of an exposure.

The RISK-UE (https://cordis.europa.eu/project/id/EVK4-CT-2000-00014) vulnerability research team, based on 1998 FEMA/NIBS categorization, adopted the four categories (pre-code, low code, moderate code, and high code) classification scheme of seismic design code era. *Milutinovic and* Trendafiloski (2003) based on dates of enforcement of seismic design standards, formulated codal eras and their enforcement span for RNM.

Considering the afore stated, the exposure model of RNM and other countries succeeding SFRY should include not only the spatial migration of design seismic zones but also temporal component depending exclusively on the time when the concrete and reinforced-concrete standards were enforced. More precisely, two conceptually different SZMs have accompanied only the JUS31/81 and two CRC standards accompanied JUS39/64 and JUS31/81, Table 5.5. Comparative analysis of CRC column reinforcement details indicates that both standards are equally non-ductile. However, a more rigorous comparative analysis is needed. A particular problem for exposure and vulnerability modelers is the migration of seismic into non-seismic zones and vice versa. The national exposure model, as recommended

Table 5.5 Design and construction eras based on enforcement time of JUS, SZM, and CRC standards

Era	Enforcement span	JUS	SZM	CRC
Pre-code	< 1948	n/a		
Low code	1948–1964	JUS61/48	SZM1948	CRC46/47
Moderate code	1965–1982	JUS39/64	SZM1964	
				CRC51/71
High code	1983–2024	JUS31/81	SZM1982	
				CRC11/87
			SZM1990	
Eurocode	≥ 2020	MKC EN 1998–1	MKC EN 1998–1/ HA:2020	MKS EN 1992–1-1

by *Milutinovic and* Trendafiloski (2003), cannot capture these effects since the record of historic design seismicity is needed for each larger human settlement.

However, having summarized uniformly, all up-to-date available seismic hazard zoning maps for RNM could be the base reference for all further studies in the seismic risk domain. RNM, until this date, has not yet performed reliable seismic risk mapping on the national level, which authors address primarily to non-consistence or non-availability of comprehensive exposure model.

We believe that with the latest initiatives overtaken by the Institute of Earthquake Engineering and Engineering Seismology (IZIIS), Ss. Cyril and Methodius University in Skopje, with the assistance and the support of the National Coordinator for Implementation of the National Platform for Disaster Risk Reduction and availability of new census data (in 2021), conditions for the development of a higher resolution national exposure model would be created very soon. It is noteworthy to stress that derived conclusions apply to all countries succeeding SFRY until they have replaced SFRY regulations with the Eurocodes.

References

CRC11/87: Rulebook for the technical standards for concrete and reinforced concrete (OGoSFRY No. 11/87 of 23.02.1987) (In Macedonian)

CRC46/47: Temporary technical regulations for concrete and reinforced concrete (OGoFNRY No. 46/47 of 03.06.1947) (in Serbian)

CRC51/71: Rulebook for technical measures and conditions for concrete and reinforced concrete (OGoFNRY No. 51/71 of 18.11.1971) (In Macedonian)

МКС EN 1998–1/НА:2020: Еврокод 8: Проектирање конструкции отпорни на земјотрес – Дел 1: Општи правила, сеизмички дејствија и правила за згради - Национален анекс / Eurocode 8: Design of structures for earthquake resistance—Part 1: General rules, seismic actions and rules for buildings—National Annex

FEMA/NIBS, HAZUS (1998) Earthquake loss estimation methodology, vol 1

Harihara Venkata Nagasai P, Bhargav Reddy V, Rama Krishna Kolli R, Nagarathinam L (2019) Comparison between seismic and nonseismic analysis of multistory building. Int J Recent Technol Eng (IJRTE) 7(6C2), ISSN: 2277-3878

IS 1893-1 (2002) Criteria for earthquake resistant design of structures, part 1: general provisions and buildings [CED 39: Earthquake Engineering]

IS 875-1 (1987) Code of practice for design loads (other than earthquake) for buildings and structures, part 1: dead loads—unit weights of building material and stored materials (Incorporating IS 1911: 967) [CED 37: Structural Safety]

JUS31/81: Rulebook on Technical Regulations for Construction of Buildings in Seismic Regions, OGoSFRY No. 31/81 of June 5, 1981 (Amendments 49/82, 29/83, 21/88, and 52/90).

JUS39/64: Rulebook on Provisional Technical Regulations for Construction in Seismic Regions, OGoSFRY No. 39/64 of September 30

JUS61/48 (1948) Provisional Technical Regulations for Loading of Structures, OGoFNRY No. 61/48 of July 17

Milutinovic ZV, Trendafiloski GS (2003) RISK-UE, An advanced approach to earthquake risk scenarios with applications to different European towns, Report to WP4: vulnerability of current buildings

Milutinovic Z, Salic R, Tomic D (2016) Seismic zoning map of the Republic of Macedonia according to the requirements of MKC EN 1998-1:2004—Eurocode 8 (map reading guide), Report IZIIS 2016–26, August 2016 (in Macedonian)
SZM1948: Division of the territory of Federal People's Republic of Yugoslavia into seismological zones, OGoFNRY No. 61/48 of July 17, 1948 (In Serbian)
SZM1964: S.F.R. Yugoslavia Seismological Map, OGoSFRY No. 39/64 of September 30, 1964
SZM1979: Seismic Zoning Map of S.R. Macedonia, OGoSRM No. 2/79 of January 31, 1979
SZM1982: Provisional Seismological Map of SFRY, OGoSFRY No. 49/82 of August 13
SZM1990: Socialist Federal Republic of Yugoslavia, Seismological Maps for Return periods of 50, 100, 200, 500, 1000 and 10000 years, OGoSFRY No. 52/90 of September 7, 1990
SZM2020: Republic of Macedonia, Seismic Zoning Map with Elements of Seismic Hazard for Return Periods of 95 and 475 years, MKC EN 1998–1/HA:2020

Chapter 6
Way Forward

Abstract This chapter outlines future directions for seismic hazard assessment globally and nationally. At the global level, it emphasizes the shift toward physics-based rupture simulations, high-resolution models, and integration of big-data analytics and machine learning into PSHA frameworks to better capture complex fault interactions and reduce epistemic uncertainty. It highlights the importance of interdisciplinary collaboration among seismologists, geologists, engineers and data scientists, and the role of initiatives such as GEM and EFEHR in providing community models, shared infrastructure and benchmarking. Enhanced real-time monitoring, early warning systems, and effective communication of hazard results to policymakers and the public are identified as key elements for operationalizing risk reduction. At the national level, the chapter presents ongoing efforts at IZIIS to revise the SZM2020 and the national annex to Eurocode 8 in line with the second-generation EC8, shifting from PGA-anchored spectra to S_α and S_β spectral ordinates and incorporating improved seismotectonic models and GMMs. It also underlines North Macedonia's active participation in EFEHR and the National Platform for Disaster Risk Reduction as a basis for strengthening resilience and integrating advanced hazard science into policy and planning.

Keywords Future seismic hazard assessment · Physics-based simulations · Machine learning · Eurocode 8 second generation · National seismic zonation · IZIIS · Disaster risk reduction and resilience

On a Global Level…

The future of seismic hazard analysis is increasingly focused on using advanced physics-based simulations and high-resolution models to understand earthquake processes better. By incorporating big data analytics and machine learning, predictions can become more accurate and help identify earthquake precursors. Improved probabilistic seismic hazard assessment (PSHA) models will include physics-based simulations to account for complex fault interactions.

R. Salic Makreska and Z. Milutinovic, *Seismic Hazard Zonation for the Republic of North Macedonia*, SpringerBriefs in Geography,
https://doi.org/10.1007/978-3-032-13945-0_6

Collaboration among seismologists, geologists, engineers, and data scientists is crucial, with community efforts like the GEM and EFEHR leading the way. Interdisciplinary approaches are becoming standard practice, ensuring that the latest scientific findings are swiftly integrated into practical applications. Enhancing real-time seismic monitoring and early warning systems provides immediate data for ongoing simulations and timely alerts. High-performance and cloud-based computing systems enable complex simulations, previously limited by computational constraints, ensuring rigorous validation and benchmarking against historical data.

Effective communication of seismic hazard results to policymakers and the public is vital for implementing risk mitigation strategies and resilience planning. Efforts are being made to translate technical data into actionable insights that can be easily understood by non-experts. Tailoring hazard models to urban environments allows detailed assessments of impacts on infrastructure and populations, guiding community resilience planning. As these methodologies advance, seismic hazard analysis will become more precise and reliable in mitigating earthquake risks. Continued innovation and investment in these areas will be crucial for adapting to the challenges posed by a changing climate and increasing urbanization.

On a National Level…

As a continuation of previous efforts in seismic hazard assessment, a team of researchers from IZIIS (Salic Makreska, Neziri, and others), with the support of external and international experts and academicians, is currently working on the revision of SZM (2020) to meet the requirements of the second generation of Eurocode 8 (date of publication: September 30, 2027). In this revision, the spectral shape will no longer be anchored at very short periods to the peak ground acceleration (PGA), but instead to two spectral ordinates: S_α (the average of a range of short-period spectral ordinates) and S_β (the 1-s spectral ordinate). The new and revised national model will include significant advancements in seismicity representation, seismotectonic modeling, and better representation of GMMs, based on advanced science, data, and analysis tools. These improvements will lead to more accurate seismic hazard maps that reflect the latest understanding of local and regional seismic activity.

Following the administrative procedures of the Standardization Institute of RNM (ISRSM), the draft of a new national annex document is expected to be submitted before the new EC8 date of publication. Recognized by the latest version of the National Platform for Disaster Risk Reduction (2019), IZIIS is one of the key institutions within the crisis management system of N. Macedonia (Chapter 6; 6.9). It is actively involved in the national efforts to build and strengthen national resilience and contributes to global efforts in disaster risk reduction.

Since September 2020, IZIIS as the leading national institution has become a voting member of the EFEHR Consortium General Assembly, with a representative in the Executive Committee since October 2023 (http://www.efehr.org/efehr/governance/). This membership allows IZIIS to stay at the forefront of seismic research and policy development, contributing to and benefiting from the collective expertise of the consortium. The aim was to join the European and global processes, follow advancements in seismic hazard and risk assessment, and contribute to their further developments. Through these collaborative efforts, IZIIS is further enhancing its capabilities and reinforcing its role as a competent institution on both national and international levels.

Last but not least, this contribution clearly illustrates the critical importance of further strengthening the interdisciplinary collaboration among various engineering fields. By integrating such holistic approach into strategic planning, we can better demonstrate the value and importance of seismic hazard assessment in informing policy and ensuring comprehensive preparedness. Furthermore, this approach highlights the necessity of combining different insights from geotechnical, structural, and environmental engineering to create robust and effective mitigation strategies.

Reference

SZM2020 (2020) Republic of Macedonia, Seismic Zoning Map with Elements of Seismic Hazard for Return Periods of 95 and 475 years, MKC EN 1998-1/HA:2020

MIX
Papier aus verantwortungsvollen Quellen
Paper from responsible sources
FSC® C105338

If you have any concerns about our products,
you can contact us on
ProductSafety@springernature.com

In case Publisher is established outside the EU,
the EU authorized representative is:
Springer Nature Customer Service Center GmbH
Europaplatz 3, 69115 Heidelberg, Germany

Printed by Libri Plureos GmbH
in Hamburg, Germany